MICROBIOLOGICAL RESEARCH – FOOD PRESERVATION AND PLANT ANTIMICROBIAL ACTIVITIES

MICROBIOLOGICAL RESEARCH – FOOD PRESERVATION AND PLANT ANTIMICROBIAL ACTIVITIES

Zakaria Ahmed

ELIVA PRESS

ISBN: 978-99949-8-353-7

Cover Design: Eliva Press
Cover Image: Ingimage
Email: info@elivapress.com
Website: www.elivapress.com

This Eliva Press imprint is published by the registered company Eliva Press Global Ltd. part of Eliva Press S.R.L. Publishing Group

The registered company address is: Pope Hennessy Street Level 2, Hennessy Tower Port Louis, Mauritius
Eliva Press S.R.L. Publishing Group address is: Bulevardul Moscova 21, Chisinau, Moldova, Europe

MICROBIOLOGICAL RESEARCH –
FOOD PRESERVATION AND PLANT ANTIMICROBIAL ACTIVITIES

ZAKARIA AHMED, Ph.D.

BANGLADESH JUTE RESEARCH INSTITUTE,
DHAKA 1207, BANGLADESH

TABLE OF CONTENT

LIST OF TABLES

LIST OF FIGURES

DEDICATE

TO

MY PARENTS

PREFACE

Foods, microorganisms and humans have had a long and interesting association that developed long before the beginning of recorded history. Wholesome, nutrient-rich foods are important for all people. Foods are not only of nutritional value to those who consume them but often are ideal culture media for microbial growth. The Food Microbiology focuses on the general biology of the microorganisms that are found in foods, although organic chemistry is also involved. Foods, because they provide nutrients for us, also are excellent environments for the growth of microorganisms. Microbial growth is controlled by factors related to the food itself (intrinsic factors), and also to the environment where the food is being stored (extrinsic factors). The intrinsic or food-related factors include pH, moisture content, water activity or availability, oxidation-reduction potential, physical structure of the food, available nutrients, and the possible presence of natural antimicrobial agents. Extrinsic or environmental factors include temperature, relative humidity, gases (CO_2, O_2) present, and the types and numbers of microorganisms present in the food. Fermentation by some microorganisms can lead to food preservation instead of food spoilage. Food spoilage is a major problem in all societies. This can occur at any point in the course of food production, transport, storage and / or preparation. Food-borne toxins are of increasing concern, especially with increases in international shipments and extended storage of food products before use. Growth of fungi can result in the synthesis of toxins such as aflatoxins, fumonisins, and ergot alkaloids. Algal-derived toxins can be transmitted to humans through freshwater and marine-derived food products. Foods can be preserved by physical, chemical, and biological processes. Refrigeration does not significantly reduce microbial populations but only retards spoilage. Pasteurization results in a pathogen-free product with a longer shelf life. Chemicals can be added to foods to control microbial growth. Such chemicals include sugar, salt (decreasing water availability), a many organic chemicals that affect specific groups of microorganisms. Microbial products, such as bacteriocins, can be added to foods to control spoilage organisms. During the entire sequence of food handling, from the producer to the final consumer, microorganisms can affect food quality and human health. Foods can transmit a wide range of diseases to humans. In food infection, the food serves as a vehicle for the transfer of the pathogen to the consumer. With food intoxication, the microorganisms grow in the food and produce toxins that can then affect the consumer. New pathogens continue to be discovered. This may be due to an increased ability to detect these pathogens as well as to better reporting and monitoring of disease occurrence. Foods that are consumed raw, such as sprouts, shellfish / finfish and fruits, are increasingly popular. There is a risk of disease transmission if care is not taken in production, storage, and transport of these perishable foods. Without adequate care, major disease outbreaks can occur because foods travel around the globe in extremely short times. Detection of food-borne pathogens is carried out using classic culture techniques, as well as immunological and molecular procedures. Modern techniques make it possible to quickly link food-borne disease occurrences to common pathogen sources. With kit-type rapid diagnostic systems, pathogens often can be detected in 1 to 2 days. The detection and control of pathogens and food spoilage microorganisms are important parts of food microbiology. Dairy products, grains, meats, fruits, and vegetables can be fermented. It has been suggested that some fermented dairy products have antimicrobial and anticancer characteristics. The making of bread, sauerkraut, sufu, pickles, and many other foods also involves the use of complex fermentation processes. When chopped plant materials are fermented, silages are created, which can be stored and used by animals. Microbial cells can be used as food sources and food amendments. These include mushrooms, cyanobacteria such as Spirulina and yeasts.

CHAPTER ONE

1.1 STUDY OF SPICES IN MEAT PRESERVATION

1.2 MICROBIOLOGICAL STATUS OF DIFFERENT SWEET AND SWEET PRODUCTS

1.3 IRRADIATION EFFECT RITA FISH KEBAB PRESERVATION

1.1 STUDY OF SPICES IN MEAT PRESERVATION

SOURCE: *Banik A, Abony M, Ahmed Z and Shishir MA. 2022. Feasibility and efficacy study of spices in meat preservation. Stamford Journal of Microbiology. 12(1): 31-36.*

1.1.1 INTRODUCTION

Spices and herbs are a few daily agents in Bengali cuisine to enhance the taste and flavor of both cooked and raw foods. The use of these spices and herbs is in fact for traditional taste which people prefer even without knowing their health benefits. The history of their use as an enhancer of flavor and aroma in food, folk medicines and food preservatives among different cultures is common and long (**1**). With progression, the potential therapeutic values including antimicrobial, antioxidant and anticancer properties were discovered (**2**). With the advent of the new era of access to information, awareness and interest grew among people of all classes the spices and herbs due to food safety concerns with increased demand for systematic application in native foodstuffs as well as scientifically optimized combinations for maximum health benefits. Consequently, many natural food preservatives were derived from spices and herbs such as cuminaldehyde from cumin, eugenol from clove, cinnamaldehyde from cinnamon, etc. which are being used to preserve many processed foods nowadays (**3**).The ability of spices, herbs, and other plant extracts to the preservation of foods is attributed to the presence of certain major bioactive compounds including phenolic acids, terpenes, aldehydes, flavonoids, and more which function as strong antioxidant and antimicrobial agents. Inhibition of microbial growth and lipid oxidation facilitates the extension of shelf-life, quality, and safety of food products (**4, 5**). Hence, the wide and frequent application of these natural preservatives could aid in reducing chemical similitude and associated chronic deleterious health effects.

In Bangladesh, varieties of spices such as onion, garlic, ginger, bay leaves, turmeric, coriander, pepper, cumin, black cardamom, cinnamon, clove, black cumin, black pepper, black seed, etc. are used mainly in different combinations forcooking different dishes. A good number of spices were reported with antagonistic activities against several food-borne pathogens in vitro as well as in the prepared food(**6–8**). Food-borne outbreaks are common here in Bangladesh due to poor hygiene practices and a favorable climate for microbial growth. These outbreaks are noticed rarely only when severe illness erupts (**9**). Appropriate hygiene practices in personal life as well as in food preparation and the optimized application of spice could be the keys toreducing such cases in Bangladesh could be the keys to reducing such cases in Bangladesh. Hence, besides improving hygiene practices, potential spices for different food items should be screened following in vitro antimicrobial assay and microbial quality assessment of foods after application.

Spice-food combination depends mostly on traditional as well as personal tastes. The cooking of eat here involves a series of spices that extends the shelf life of the item along with the taste. On the of the contrary, storage of raw meat involves different techniques such as chilling/refrigeration, freezing, brining, curing, smoking, thermal processing, canning, dehydration, irradiation, chemicals, pressure processing, etc. so that the chances of pathogenic microbial growth, oxidation, andenzymatic spoilage is reduced (**10–12**). The use of spices in the preservation of raw meat at low temperatures was reported to be successful (**13–15**). Since the technique is a simple one,it could be used by the mass population for the safe storage of their meat. The objective of this study was to evaluate the antimicrobial potential of spice extracts against food-borne bacterial pathogens and their application in beef preservation.

1.1.2MATERIALS AND METHODS

1.1.2.1 Bacterial Strains Sixpathogenic bacteria viz.*Acinetobacter* sp., *Bacillus cereus* (ATCC 14579), *Escherichia coli* (ATCC 25922), *Salmonella enterica* ser. *typhi* (ATCC 14028), *Staphylococcus aureus* (ATCC 6538) and *Vibrio cholerae* (ATCC 14035) obtained from the Department of Microbiology, Primeasia University, Bangladesh, were stocked in a medium containing 20% glycerol in cryogenic vials at -70°C. Working cultures were maintained on Tryptic Soy Agar (TSA) slants at 4°C and fresh slants at specific time intervals were used for subculture.

1.1.2.2 Preparation of Ethanol Extracts from Spices: Seven common spices used in this study were black cumin (*Nigella sative*), black pepper (*Piper nigrum*), garlic (*Allium sativum*), ginger (*Zingiber officinale*), cumin (*Cuminum cyminum*), mustard (*Brassica nigra*) and onion (*Allium cepa*). Fresh bulbs of garlic and onion, the fresh root of ginger, fresh dried fruits of black pepper, and the fresh seeds of cumin, mustard, and black cumin were cleaned, washed with sterile distilled water, sliced, and dried on trays in a hot air oven at 50°C the brittleness. The dried spices were then ground in a sanitized blender and 20g of each spice powder was added aseptically into 80ml of 95% ethanol and allowed to be soaked overnight at room temperature in a reciprocal shaking platform at 150rpm. The separation of the ethanol fraction was done with a sterilized cheesecloth and filtered subsequently through a sterilized Whatman filter paper (No. 3). The ethanol fraction was then kept in an oven (Binder,USA) at 40°C to evaporate any residual ethanol. The extract was weighed and dissolved in ethanol to a concentration of 10mg/ml (**16**), and the dried substance was kept in a sterile bottle under refrigerated condition for further use.

1.1.2.3 Inoculum preparation The inocula of the test organisms were prepared by transferring one loop-full of the colony from TSA (Sigma, USA) medium into 9.0 ml of sterile Mueller-Hinton Broth (Difco, Sparks, MD) and incubated at 37°C for 5 to 6h. The bacterial cultures were compared with the McFarland turbidity standard (10^8 cfu/ml) prepared by mixing 0.05ml of 1.175% Barium chloride dihydrate ($BaCl_2.2H_2O$), with 9.95ml of 1% sulfuric acid (H_2SO_4) in a test tube with constant stirring (**16, 17**).

1.1.2.4 Agar well diffusionassay The antibacterial activity of ethanol extracts of spices was evaluated following the agar well diffusion method (**18**). In brief, to prepare bacterial lawn of test pathogens, overnight liquid cultures in MHB medium at desired turbidity of 0.5M McFarland standard were seeded on Muller Hinton Agar (MHA) plates with sterilized cotton swabs (**16, 17**). Thecotton swabs were soaked in the liquid cultures and applied evenly in three directions at an angle of 60°C onto the surface of MHA plates. A sterile borer was then used to make 5 wells of 8.0mm diameter in each plate, once the wet surface of the MHA medium was dried up. Ethanol extracts of the spices were poured separately in amounts of 100μl into the wells along with negative control, the very ethanol used in the extraction process. After 12-15min of diffusion time at room temperature, the plates were incubated at 37°C for 48h and the zone of inhibition (ZOI) was measured in 3 directions and recorded for the ethanol extracts. For a certain spice against pathogens, the diameter of the ZOIs more than 12mm was considered as inhibition (**19**), diameters between 12-16mm were considered moderately active, and greater than 16mm were considered highly active. Each experiment was performed in triplicate and mean values were taken.

1.1.2.5 Determination of MIC and MBC The MIC and MBC of the crude extracts were determined following the tube dilution techniques with MHB as diluents and growth medium (**16**). In brief, 1.0ml of the reconstituted ethanol extract (512μg/ml) was added to 1.0ml of MHB medium, and simulation of the fashion helped in obtaining a series of concentrations i.e. 512, 256, 128, 64, 32, 16 and 8 (μg/ml). Besides, the turbidity of the overnight grown test pathogens on MHB medium was adjusted to 0.5M McFarland standard to be used as inoculums (**16**). Then, 100μl of each inoculum was mixed well with the serially diluted crude spice extracts, and the vials retaining the mixtures were incubated at 37°C for 24h. A vial devoid of any crude spice extract but pure diluent (1.0ml) was kept as a negative control. The lowest concentrations of the extracts hardly demonstrating any visible growth were considered the MICs. One loop-full of the mixture from the MIC tubes was sub-cultured on TSA plates and incubated overnight at 37°C and the MICs yielding no growth were recorded as the MBCs.

1.1.2.6 Sample preparation The beef sample was obtained from a local butcher shop in Dhaka and collected into insulated polystyrene boxes placed in ice and transported to the laboratory immediately usingthe standard microbiological procedure. The beef was washed with sterile saline in the bio-safety cabinet and cut into 50g pieces with a sterile knife. The beef pieces were then soaked into the solutions of selected treatments i.e. 0.85%NaCl, vinegar, and separate solutions of extracts of all spices prepared in sterile distilled water. Raw beef pieces were also kept as no treatment. Beef pieces were then transferred into sterile containers and stored at room temperature for 3 days for further microbiological assessment. The experiment was performed in triplicates.

1.1.2.7 Microbiological analysis The influence of the spice extracts on the microbiological quality of the beef was determined by estimating the Total Viable Count (TVC) of the treated beef samples after 72h of incubation at room temperature. In this connection, the beef pieces were transferred aseptically into separate stomacher bags containing 90ml of sterile buffered peptone water solution and homogenized properly for 2min. The beef homogenates were then diluted serially from 10^{-1} -10^{-8}folds in 0.85% NaCl solutions and 100µl of each dilution was spread on the Nutrient agar plate. After overnight incubation at 37°C, all the plates were checked and the numbers of colonies were recorded. The numbers were then expressed as log 10 cfu/g of beef.

1.1.3RESULTS

1.1.3.1 Antibacterial activity of testedspices The antibacterial assay revealed that the ethanol extracts of spices used in this study potentially inhibited the test organisms at varied levels (**Table 1.1.1**). Among them, *B. cereus* (19.57±8.05 mm) was the mostly inhibited bacterium followed by*Acinetobacter* sp. (18.14±1.34 mm), *E. coli* (16.28±1.88mm), *S. aureus* (14.28±9.91mm), *V. cholerae* (10.85±7.94mm) and *Salmonella enterica* ser. *typhi* (6.85±8.55mm) by the spice extracts. Based on the inhibitory activity against the test organisms, the most active spice was garlic (17.83±2.48) followed by black pepper (17±8.9), black cumin (15.83±10.87), ginger (15.16±7.68), cumin (14±7.5), mustard (10.5±8.47) and onion (10±8.19). Garlic was most effective against *S. aureus* (22mm) followed by*Acinetobacter* sp. (19mm), *E. coli* (18mm), *B. cereus* (17mm), *S.*

enterica ser. *typhi* (16mm), and *V. cholerae* (15mm). Garlic also demonstrated the most consistent inhibitory activity (sd- ±2.48) against all organisms although the black cumin exhibited the highest activity (ZOI= 34 mm) against *B. cereus* whereas garlic was moderately active. The inhibitory level of black pepper was very close to garlic. Nevertheless, black pepper was ineffective against *S. enterica*ser. Typhi (no inhibition at all), caused maximum inhibition to *E. coli* (18mm) and *V. cholerae* (21mm). *S. aureus*and *Acinetobacter* sp. were also moderately inhibited by Black pepper. Black cumin, ginger, and cumin inhibited the organism at a moderate level. Interestingly, *S. enterica* ser. Typhi was inhibited equally (16mm) by ginger, cumin, and garlic whereas black cumin, black pepper, mustard, and onion were not at all inhibiting. Cumin and garlic inhibited *S. aureus* maximum (18mm) but onion and mustard did notexhibit any inhibitory activity. Onion inhibited *E. coli*maximum (18mm) besides black pepper and garlic. All of the antibiotics used in this study as controls demonstrated strong bacteriocidal activity except AP10, NA30, and NI300. *Acinetobacter* sp. was found to be completely resistant against AP10 and NA30 and *V. cholerae* was completely resistant against NA30 and NI300. The average diameters of zones of inhibition against the test organisms by AP10 as well as NI300 were around 15 mm and by NA30, it was 11.16±8.9. Again, *Acinetobacter* sp. and *V. cholerae* demonstrated resistance against these antibiotics to some extent (**Table 1.1.1**).

Table 1.1.1: Antibacterial potentials of tested spices and commercial antibiotics [Banik et al 2022]

Agents	*B. cereus*	*E. coli*	*Acinetobacter*	*S. aureus*	*V. cholerae*	*Salmonella*	Mean±SD
Black Cumin	34	13	16	17	15	0	15.83±10.87
Black pepper	26	18	17	20	21	0	17±8.9
Ginger	21	16	19	19	0	16	15.16±7.68
Onion	12	18	18	0	12	0	10±8.19
Mustard	14	16	20	0	13	0	10.5±8.47
Cumin	13	15	18	22	0	16	14±7.5
Garlic	17	18	19	22	15	16	17.83±2.48
Mean±SD	19.57±8.05	16.28±1.88	18.14±1.34	14.28±9.91	10.85±7.94	6.85±8.55	
GEN10	24	22	20	22	23	14	20.83±3.6
Lev 5	22	26	16	22	25	20	21.83±3.6
AP 10	15	21	0	18	16	20	15±7.6
IPM 10	26	21	18	22	10	25	20.33±5.8
CIP 5	23	27	17	22	22	15	21±4.3
NA 30	13	20	0	16	0	18	11.16±8.9
MEM 10	30	25	22	28	22	18	24.16±4.4
NI 300	24	20	18	18	0	14	15.66±8.33
Mean±SD	22.12±5.59	22.75±2.81	13.87±8.75	21±3.7	14.75±10.26	18±3.74	

Note: Gentamicin (GEN 10µg), Levofloxacin (LEV 5µg), Ampicillin (AP 10µg), Imipenem (IPM 10µg), Ciprofloxacin (CIP 5µg), Nalidixic acid (NA 30µg), Meropenem (MEM 10µg), Nitrofurantoin (NI 300µg)

1.1.3.2 MIC and MBC of spice extracts The sensitive bacteria, as revealed through the antibacterial assay of ethanol extract of spices, were further exposed to different dosages comprised of gradient concentrations for the determination of MIC and MBC. The minimum inhibitory concentration (MIC) and the minimum bactericidal concentration (MBC) of the spices against the test organisms were determined thus. It was found that the MIC and MBC of ethanol extracts of spices against B. cereus were in the range of (32-256mg/ml) and (64-512mg/ml) respectively. In both cases, black cumin was found to be the most efficient with a minimum of 32mg/ml (MIC) and 64mg/ml (MBC) whereas black pepper, ginger, and garlic exerted the same at a slightly higher dose (MIC-64mg/ml and MBC-128mg/ml) (**Table 1.1.2**). For *E. coli*, MIC and MBC were in the range of (64-256mg/ml) and (128-512mg/ml), respectively and the lowest concentrations were recorded for garlic (MIC-64mg/ml and MBC- 128mg/ml). For *Acinetobacter* sp., MIC and MBC were in the range of(64-128mg/ml) and (128-256mg/ml) respectively and the inhibitory dosages were almost the same for black pepper,onion, mustard, garlic, and cumin i.e. MIC-64 mg/ml and MBC-128mg/ml. For *S. aureus*, MIC and MBC were in the range of (64-128mg/ml) and (128-256mg/ml) respectively and the lowest dosages were observed for garlic and cumin (MIC-64mg/ml and MBC-128mg/ml). For *V. cholerae*, MIC and MBC were in the range of (64-256mg/ml) and (128-512mg/ml) respectively and the lowest dose was recorded for black pepper. For *Salmonella* ser. Typhi, MIC and MBC were in the range of (128-256mg/ml) and (256-512mg/ml), respectively with garlic producing the most inhibitory effect (MIC-128mg/ml and MBC-256mg/ml). The spices, earlier proven to be inefficient, were not considered in this case and hence remarked with asterisks (*: not performed).

Table 1.1.2: MIC and MBC of extracts (mg/ml) of all spices against foodborne bacterial pathogens[Banik et al 2022]

Organisms	Black cumin		Black pepper		Ginger		Onion		Mustard		Garlic		Cumin	
	MIC	MBC	MIC	MBC	MIC	MBC	MIC	MBC	MIC	MBC	MIC	MBC	MIC	MBC
B. cereus	32	64	64	128	64	128	256	512	256	512	64	128	256	512
E. coli	256	512	128	256	256	256	128	256	128	256	64	128	256	512
Acinetobacter sp	128	256	64	128	128	256	64	128	64	128	64	128	64	128
S. aureus	128	256	128	256	128	256	*	*	*	*	64	128	64	128
V. cholerae	128	256	64	128	*	*	256	512	256	512	128	256	*	*
S. enterica	*	*	*	*	256	256	*	*	*	*	128	256	256	512

Note: * =not performed.

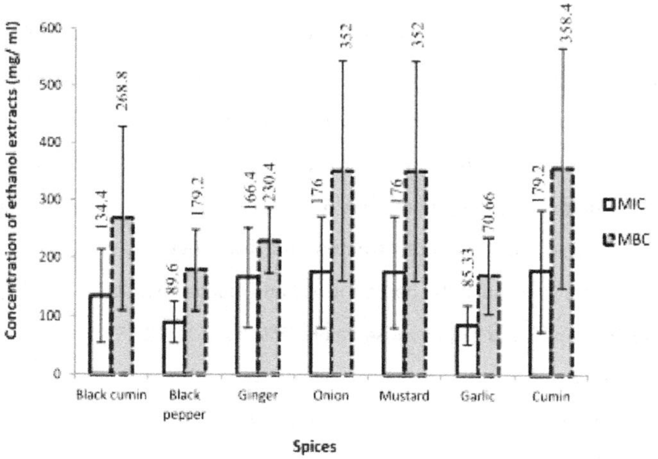

Figure 1.1.1: Average MIC and MBC of ethanol extracts of spices against the tested food-borne pathogens. [**Banik et al 2022**]While considering the overall inhibitory effects of ethanol extracts of spices against the test organisms, garlic was found to be the best candidate due to its lowest average MIC (85.33±33 mg/ml) and MBC (170.66±66 mg/ml) and inhibited all the pathogens consistently. Black pepper (89.6±35 mg/ml) was very close to garlic and black cumin was also found to produce higher inhibitory effects against the test organisms. But black pepper and black cumin failed in inhibiting *S. enterica* ser. Typhi.

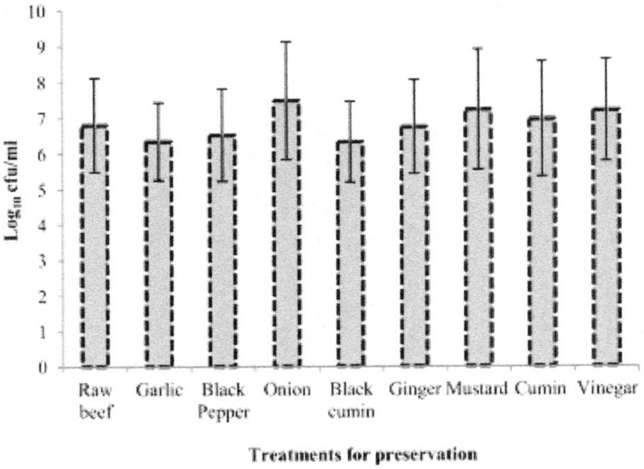

Figure 1.1.2: Effect of spice extracts on the microbial load of raw beef after 72 hours of treatment. [Banik et al 2022]Treating raw meat with spice extracts, especially garlic, black pepper, black cumin and ginger was found to inhibit microbial growth.

The efficiency of spice extracts in beef preservation: The total viable count (TVC) of raw beef, as estimated before any treatment, was 6.55×10^6 cfu/ml. After 72 hours of incubation at 37°C, the microbiological quality of both treated and raw meat was compared. It was revealed that the spice extracts exerted both bacteriocidal and bacteriostatic effects (**Fig. 1.1.2**). Black cumin, garlic, black pepper, and ginger caused a 66.66%, 65.55%, 48.18%, and 11.93% reduction in the total viable count of the treated beef respectively whereas the microbial growth was enhanced in the untreated raw beef as revealed from the TVC which wastoo numerous to count (TNTC). Although cumin did not reduce the TVC, it hampered the normal multiplication of the microorganisms, thus causing bacteriostatic effects. On the other hand, onion, mustard, and vinegar retarded the growth rate of the microorganisms.

1.1.4DISCUSSION

Food could be contaminated with pathogenic microorganisms through a series of natural events and artifacts in its journey from farm to fork. Thechallenges with both pathogenic and non-pathogenic microbes are being augmented worldwide with times since multi-drug resistance and the transfer of these antibiotic resistance genes are reaching beyond control. Uncontrolled, unnecessary, and sometimes un-prescribed use of antibiotics is causing such damage which should be recovered in a holistic approach. Both reducing unnecessary useof antibiotics and ensuring safe foods devoid of pathogenic microbes should be implemented simultaneously besides other practices. The access of pathogens could be kept limited by sanitizing fruits and vegetables before consumption or the microbial load must be maintained below the thresholds for long-term storage. The latter is usually achieved with chemical preservatives which very often cause chronic health complexities and their substitution is highly recommended. The study was, therefore, aimed at the assessment of the antimicrobial potentials of indigenous spices and their feasibility as natural food preservatives. The common kitchen spices were selected in this study and their ethanol extracts were tested against 6 important food-borne pathogenic bacteria. Since the ethanol fraction of herbs and spices was reported to contain potential anti-microbial compounds such as terpenes and terpenoids (**16, 20**), this very fraction of spices was prepared and tested. It was observed from the study that the ethanol extracts of the spices potentially inhibited the test organisms at a varied level. The test organisms were comprised of 2 Gram-positive bacteria (*B. cereus* and *S. aureus*) and 4 gram-negative bacteria (*Acinetobacter* sp., *E. coli, S. enterica* ser. Typhi, and *V. cholerae*). It is well known that spices are more active against gram-positive bacteria than gram-negative ones (**7, 21**). Here in this study, although *B. cereus* (19.57±8.05 mm) was the mostly inhibited bacterium, *Acinetobacter* sp. (18.14±1.34mm), *E. coli*(16.28±1.88mm) were more susceptible to the spices than gram-positive *S. aureus* (14.28±9.91mm). It could therefore be an interesting finding since spices were exerting inhibitory activity against Gram-negative bacteria. More interestingly, the average bioactivity of spices against *Acinetobacter* sp. was 18.14±1.34mm (ZOI) whereas 23.5% lower efficacy (13.87±8.75) was observed for the commercial antibiotics. Again, the spices were more consistent in inhibiting the *Acinetobacter* sp. than the commercial antibiotics. This finding is in accordance with the report of Rath et al. where the *Acinetobacter baumannii* was found to be resistant against almost all tested antibiotics but sensitive against the methanolic extracts of spices (21.81±3.56) (**22**). On the contrary,*S. enterica* ser. Typhi (6.85±8.55) and *Vibrio cholerae* (10.85±7.94) were found to be less inhibited than other 4 test organisms by the spice extracts. Their sensitivity against the antibiotics was also moderate.

While compared to commercial antibiotics, the tested spices extracts showed convincing antibacterial potentials (**Table1.1.1**). Meropenem (MEM 10µg), Imipenem (IPM 10µg), Ciprofloxacin (CIP 5µg), Gentamicin (GEN 10µg), and Levofloxacin (Lev 5µg) (Oxoid, UK) were found to be more active than thespices but not exceeding 35.5%. Meropenem (MEM 10) demonstrated maximum activity (35.5% more than that of garlic) followed by Lev 5 (22.4% more), CIP 5 (17.78% more), GEN10 (16.8% more), IPM 10 (14% more). Since black cumin demonstrated maximum antibacterial activity (34mm) against *B. cereus*, even more than that of MEM (30mm), this spice could be the source of a very efficient and safe food preservative. This resembled other study reports that Black cumin strongly inhibited *B. cereus*(**6, 23**). Apprehending cases were the complete resistance of *Acinetobacter* sp. against Ampicillin (AP10) and Nalidixic acid (NA30), and *V. cholerae* against NA30 and Nitrofurantoin (NI 300). Fortunately,*Acinetobacter* sp. was inhibited by all spices which suggest that spices could be, besides their uses as food preservatives, the sources of efficient antimicrobial drugs, and *V. cholerae* was inhibited most by black pepper followed by black cumin and garlic. Hence, for instance, Black cumin highly active against *B. cereus* (ZOI: 34mm), should be analyzed more to develop efficient and safe food preservatives to ward off bacillary dysentery. Garlic extract demonstrated satisfactory inhibition (17.83±2.48) against all tested organismseven *S. aureus, V. cholerae, S.* ser. *Typhi*which suggests that garlic should be a must spice in the dish or as a salad dressing item and effective drugs could be developed from it.Then, MIC and MBC of those ethanol extracts were estimated. Garlic demonstrated a consistent inhibition of all test organisms at a low concentration (mean value: 85.33mg/ml) followed by black pepper (mean value: 89.6mg/ml). Among the spices, only garlic was pepper which were 6.35, 6.34, and 6.53 log cfu/gm (**Fig. 1.1.2**), respectively. On the other hand, four spice extracts (cumin, mustard, ginger, and onion) treatments resulted in a slow increase of TVC samples after 72h. The vinegar treatment resulted in higher TVC in the beef after 72h. The spices, especially garlic, black pepper, and black cumin were found to possess satisfactory levels of broad-spectrum antimicrobial activities and could be utilized as safe and effective natural food preservatives.

1.1.5CONCLUSION

The efficacy of the spices, especially garlic, black pepper, and black cumin, in inhibiting food-borne pathogens as well as in beef preservation could be observed in this study. Since few antibiotics were found to be inefficient against certain pathogens whereas spice extracts inhibited them successfully, more attention should be on the systematic use of spices in food preparation rather than the frequent use of antibiotics.

1.1.6REFERENCES

1. Hyldgaard M, Mygind T and Meyer RL. 2012. Essential Oils in Food Preservation: Mode of Action, Synergies, and Interactions with Food Matrix Components. Front. Microbiol. 3: 12.

2. Gottardi D, Bukvicki D, Prasad S and Tyagi AK. 2016. Beneficial effects of spices in food preservation and safety. Front. Microbiol. 7: 1-20.

3. Jessica Elizabeth DLT, Gassara F, Kouassi AP, Brar SK and Belkacemi K. 2017. Spice use in food: Properties and benefits. Crit. Rev. Food Sci. Nutr. 57(6): 1078-88.

4. Aziz M and Karboune S. 2016. Natural antimicrobial/antioxidant agents in meat and poultry products as well as fruits and vegetables: A review. Crit. Rev. Food Sci. Nutr. 20: 1-26.

5. Shelef LA, Naglik OA and Bogen DW. 1980. Sensitivity of Some Common Food-borne Bacteria to the Spices Sage, Rosemary, and Allspice. J. Food Sci. 45(4): 1042-4.

6. Abdel-Raouf M, Nabil M and El-Sayed M. 2014. Antimicrobial activities of some herbs extracts on food borne bacteria. J. Am. Sci. 10(11): 76-85.

7. Ağaoğlu S, Dostbil N and Alemdar S. 2007. Antimicrobial activity of some spices used in the meat industry. Bull. Vet. Inst. Pulawy. 51(1): 53-8.

8. Dyankova S, Dyakova A, Miteva D and Garbacheva M. 2009. Study of antibacterial activity of some spice extracts used in the meat industry. Agric. Sci. Technol. 1(Pilot): 43-7.

9. Noor R and Feroz F. 2017. Food safety in Bangladesh: A microbiological perspective. Stamford J. Microbiol. 6(1): 1-6.

10. Pal M and Devrani M. 2018. Application of Various Techniques for Meat Preservation. J. Exp. Food Chem. 04(01): 1-6.

11. Banik A, Abony M, Zerin T and Datta S. 2018. Antibacterial Activity of Allium Sativum, Syzygium Aromaticum,and Cinnamomum Zeylanicum against Food Borne Pathogens in vitro. IOSR J. Pharm. Biol. Sci. 13(2): 68-73.

12. Nahor U and Ahmed Z. 2012. Antimicrobial Activity of Phyllanthus Emblica and Allium Sativum: Comparative Analysis of Antimicrobial Action of Crude and Ethanolic Extract of These Natural Plant Products. IOSR J. Pharm. Biol. Sci. 4(3): 21-6.

13. Zhang H, Wu J and Guo X. 2016. Effects of antimicrobial and antioxidant activities of spice extracts on raw chicken meat quality. Food. Sci. Hum. Wellness. 5(1): 39-48.

14. Radha KK, Babuskin S, Azhagu SBP, Sasikala M, Sabina K, Archana G et al. 2014. Antimicrobial and antioxidant effects of spice extracts on the shelf life extension of raw chicken meat. Int. J. Food Microbiol. 171: 32-40.

15. Radha KK, Babuskin S, Babu PAS, Fayidh MA, Sabina K, Archana G et al. 2014. Bio protection and preservation of raw beef meat using pungent aromatic plant substances. J. Sci. Food Agric. 94(12): 2456-63.

16. Hoque MM, Rattila S, Shishir MA, Bari ML, Inatsu Y and Kawamoto S. 2012. Antibacterial Activity of Ethanol Extract of Betel Leaf (Piper betle L.) Against Some Food Borne Pathogens. Bangladesh J. Microbiol. 28(2): 58-63.

17. Shishir MA, Mamun M Al, Mian MM, Ferdous UT, Akter NJ, Suravi RS et al. Prevalence of Vibrio cholerae in Coastal Alternative Supplies of Drinking Water and Association with Bacillus-Like Spore Formers. Front. Public. Heal. 6: 50.

18. Oke F, Aslim B, Ozturk S and Altundag S. 2009. Essential oil composition, antimicrobial and antioxidant activities of Satureja cuneifolia Ten. Food Chem. 112(4): 874-9.

19. Arora DS and Kaur J. 1999. Antimicrobial activity of spices. Int. J. Antimicrob. Agents. 12(3):257-62.

20. Cowan MM. 1999. Plant Products as Antimicrobial Agents. Clin. Microbiol. Rev. 12(4):564-82.

21. Hoque MM, Bari ML, Juneja VK and Kawamoto S. 2007. Antimicrobial Activity of Cloves and Cinnamon Extracts against Food Borne Pathogens and Spoilage bacteria, and Inactivation of Listeria monocytogenes in Ground Chicken meat with their Essential oils. Rep. Natl. Food Res. Inst. 72:9-21.

22. Rath S and Padhy RN. 2014. Monitoring in vitro antibacterial efficacy of 26 Indian spices against multidrug resistant urinary tract infecting bacteria. Integr. Med. Res. 3(3):133-41.

23. Singh G, Marimuthu P, Murali HS and Bawa AS. 2005. Antioxidative and Antibacterial Potentials of Essential Oils and Extracts Isolated From Various Spice Materials. J. Food Saf. 25(2):130-45.

24. Kedzia A. 2010. Antimicrobial activity of garlic (Allium sativum L.). Postep. Fitoter. 11(1):46-52.

1.2 MICROBIOLOGICAL STATUS OF DIFFERENT SWEET AND SWEET PRODUCTS

SOURCE:*Mesbah Uddin Md. Rezwanun Nayem and Zakaria Ahmed. 2013. Microbiological status of different sweet and sweet products in Dhaka city, Bangladesh. International Journal of Bio-Technology and Research (IJBTR). 3(1): 1-10.*

1.2.1 INTRODUCTION

The sweetmeats are delicious, wholesome, nutritious and very fame items in Bangladesh especially in Dhaka city. Sweets and sweet products occupied a significant role in each sphere of our society. On occasions like birthday, marriages, religious festivals and guest entertainment sweet products are inevitable. Different types of famous sweetmeats are available in different areas of Bangladesh. The various sweet products available in the markets are Channa based milk products. It constitutes important nutritional components for all age groups. Sweet products supply the most essential elements like Proteins, Vitamin-B, Calcium and Phosphorusalong with numerous other essential major and minor substances (**1**). Among famous sweetmeats, Rasmalai, Misti Dahi, Chamcham, Kachhagolla, Sandesh and Cream Toast are describable. The origin of Kachhagolla is claimed to be the oldest (**2**). It is very nutritious on account of its fairly high protein and fat content, minerals, specially calcium and phosphorus and also fat soluble vitamins particularly vitamine A and D (**3**). Yoghurt is locally called Dahi is the good source of Vitamin-B, Proteins and Calcium which are easier for the body to digest than when they are present in fresh milk. It contains 85-88% water, 5-8% fat, 3.2-3.4% protein, 4.6-5.2% lactose, 0.5-1.1% lactic acid, 0.70-0.75% ash, 0.12-0.14% Calcium and 0.09-0.1% Phosphorus. For some individuals, Dahi has definite therapeutic value, especially who usually suffer from stomach and intestinal disorders. The use based on the assumption that the acid fermenting bacteria and lactose of milk are able to create conditions in the intestinal tract which are unfavorable for the growth of putrefactive bacteria and thereby prevents the formation of gas and a condition known as "autointoxication" (**4**). Yoghurt is also effective in lowering the blood cholesterol. Traditionally, yoghurt is made from milk by reducing its volume. By reducing volume of milk producers got loss economically. Heat also facilitates to loss of nutrients like volatilefatty acids, vitamin C and some fatsoluble vitamins through evaporation. They will be economically benefited without reducing volume of milk. Some researches works have been done in different countries on the quality of Dahi. But very limited works have been done on the microbiological status of Dahi. Hence, an attempt was made to judge the microbiological status of Dahi. This experiment is carrying significance in this country condition because the people of our country have no idea about the new food product. This new food product can bring significant change in our national health, so this experiment is very important for practical point of view.

Rasmalai is a popular South-Asian dessert, common to India, Pakistan and Bangladesh. Rasmalai consists of sugary, cream to yellow-colored balls (or flattened balls) of paneer soaked in malai (clotted cream). Rasmalai is likely to have originated in Eastern India, especially in Orissa and Bangladesh. It containsTotal Fat 31%, Cholesterol 23%, Sodium 4%, Potassium 10%, Total Carbohydrate 16%, Dietary Fiber 2%, Protein 16%, Vitamin A 14%, Calcium 27% Iron 1%, Vitamin E 2%, Vitamin B_6 4%, Vitamin B_{12} 16% and Magnesium 7%. Chamcham is another delicious sweet product in Bangladesh. Nutritional Facts for Chamcham contains Sugars 61%, Total Carbohydrate 15%, and Calories from Fat 19%, Cholesterol 6%, Sodium 4%, and Saturated Fat 16% per serving of daily value. Sandesh is one of the modified sweet milk product. It is claimed to the very good source of carbohydrate, Protein and fat. The total fat in sandesh is 21%, Saturated Fat 20%, Cholesterol 3%, Total carbohydrate 11%, Sugars 30g and Protein 14%. Though milk is supposed to constitute a complex ecosystem for various microorganisms including bacteria. Sweet milk products like above are widely consumed and marketed for them has existed in many parts of the world for many generations. There is an increase demand by the consumer for high quality natural food, free from artificial preservatives and contaminating microorganisms (**5**).

The manufacture of sweets and sweet products is based on traditional method without any regard to the quality of raw materials used and / or the hygienic quality of the products. Under such conditions, many microorganisms can find access to the sweet and sweet products. Due to lack of quality control at the production units, lead to contamination of sweet and sweet products with various types of microorganisms through processing, handling and unhygienic conditions leading to a low shelf life of the base products. Most of these products are sold inthe market without proper packaging and unduly exposing them to atmospheric microbial contamination. For this fact, the isolation, identification and enumeration of microorganisms in different sweet and sweet products is a unique part of any good quality assurance program and reflect the effectiveness of hygienic practices, processing and distribution of local manufactures. Therefore, the present study was taken to assess the microbiological status of different sweet and sweet products in Dhaka City, Bangladesh.

1.2.2MATERIALS AND METHODS

Three different areas of Dhaka city were selected as sampling sites for investigation according to consumer's affordability such as Dhanmondi, Dhaka Cantonment and Mirpur,

Bangladesh. Sterile containers were used for sample collection. All equipment's and glass ware were sterilized by heating into Hot Air Oven for 180^{0}C for two hours. All growth medium and solutions were sterilized by autoclaving at 121^{0}C and 15lbs pressure per square inch for 20min in the autoclave. Each sample were marked properly immediately after collection and transported to laboratory. The samples were collected by the representatives of individuals whose hand had been washed and were tested promptly as possible after collection. Contaminations of samples were avoided by followingaseptic technique. All the samples were serially diluted using ten-fold serial dilution technique up to $10^{-7}$and plating techniques were used for the determination of microbial load. The plates were incubated at 37^{0}C under aseptic condition in incubator and the colonies were counted after 24h incubation. The totals viable bacterial count (TVBC) was carried out from the both pour and spread plate. Gram staining was performed by standard method and thecharacteristics (shape, arrangement and staining character) of Gram stain of isolate were observed on the microscopic field (**6**). Different types of typical colonies were screened out from the plates for morphological and biochemical test in order to determine the probable microbes.

1.2.3RESULTS AND DISCUSSIONS

The Total Viable Count (TVC) of Chamcham and Kachagolla samples showed higher (5.4×10^5cfu/g and 3.4×10^5cfu/g; respectively) in Bhayzakul brand, in sandesh samples higher 8.5×10^6cfu/g TVC was found in Muslim brand. In sweet products, higher TVC (5.6×10^2cfu/g) was found in Rasmalai. Higher TVC in Sweet Yoghurt was 8.3×10^4cfu/g in Bikrampur brand and in Cream Toast, 5.8×10^5cfu/g was observed in Bhayazakul brand. This may be due to poor quality of raw milk as pointed out by Boor et al (**7**) who reported that the bacterial quality of raw milk is increasingly important to final product quality (**Table 1.2.1**).

Table 1.2.1: Total Microbiological Status of Sweets from Different Sources. [Mesbah et al 2013]

SWEETS SAMPLES			
	Muslim Sweets	**Bikrampur**	**Bhayzakul**
Chamcham	3.5×10^4cfu/g	2.24×10^5cfu/g	5.4×10^5cfu/g
Kachhagolla	2.2×10^6cfu/g	1.93×10^6cfu/g	3.4×10^5cfu/g
Sandesh	8.5×10^6cfu/g	6.4×10^6cfu/g	5.8×10^5 cfu/g
SWEET PRODUCTS			
Rasmalai	5.6×10^2cfu/g	1.5×10^6cfu/g	1.64×10^5cfu/g
Sweet Yoghurt	5.0×10^4cfu/g	8.3×10^4cfu/g	7.6×10^4cfu/g
Cream Toast	2.4×10^6cfu/g	2.4×10^6cfu/g	5.8×10^5cfu/g

Kachhagolla is a Chhana based sweetmeat, it is very nutritious on account of its fairly high protein and fat content, minerals, specially calcium and phosphorus and also fat soluble vitamins particularly vitamin A and D. A part of produced milk used for the preparation of chhana and finally for sweetmeats especially for Kachhagolla making. There is no set standards and regulation for production of quality Kachhagolla. Although the actual preparation and standards depend on manufacturing practices of the experienced producers. The number of total viable count per gram of samples were $3.5x10^4$, $8.5x10^4$, $1.20x10^4$, $1.50x10^4$, $8.8x10^4$ and $2.20x10^4$ respectively different types of kachhagolla sample. Statistically it was found that there was a significant difference ($P<0.01$) within the total viable bacteria of different types of kachhagolla sample (**3**).

Table 1.2.2: Cultural and Biochemical Characteristics of Unknown Organism. [Mesbah et al 2013]

Name of the Sample	Name of the Source	Nature of the Colony	Gram Stain	Fermentation		H_2S Production	Indole Production	MR Reaction	VP Reaction	Citrate Reaction	Urease Activity	Catalase Activity	Name of The Probable Organism
				Lactose	Sucrose								
Rasmalai	Muslim Sweets	Cream	Rod; -ve	-	+	-	+	+	-	+	+	+	*Proteus vulgaris*
		Yellow	Cocci; +ve	+	-	-	+	+	-	-	-	+	*Streptococcus lactis*
		Pin point	Rod; - ve	+	-	-	+	+	-	-	-	+	*E. coli*
		Orange	Rod; - ve	+	-	-	-	+	-	+	+	+	*Klebsiella pneumoniae*
	Bikrampur	White Pin Point	Rod; - ve	+	-	-	-	+	-	+	+	+	*K. pneumoniae*
		White Pin Head	Rod; - ve	+	-	-	-	+	-	+	+	+	*K. pneumoniae*
		Translucent Round	Rod; - ve	+	-	-	-	+	-	+	+	+	*K. pneumoniae*
	Bhayzokul	Orange	Rod; - ve	-	+	-	+	+	-	+	+	+	*Proteus vulgaris*
		Pin Head Transperent	Rod; - ve	-	+	-	+	+	-	+	+	+	*P vulgaris*
		Pin Head White	Rod; - ve	-	+	-	+	+	-	+	+	+	*P. vulgaris*
Sweet Yoghurt	Muslim Sweets	Cream	Rod; -ve	+	-	-	-	+	-	+	+	+	*K. pneumoniae*
		White	Rod; -ve	-	+	-	+	+	-	-	-	+	*E. coli*
	Bikram pur	White Pin Point	Cocci; +ve	+	-	-	-	+	-	-	-	-	*S. lactis*
	Bhayzokul	White Pin Point	Cocci; +ve	+	-	-	-	+	-	-	-	-	*S. lactis*

Cream Toast	Muslim Sweets	White Pin Head	Rod; -ve	+	-	-	-	+	-	+	+	+	*K. pneumoniae*
		Cream	Rod; -ve	+	-	-	-	+	-	+	+	+	*K. pneumoniae*
	Bikram pur	White Dispersed	Rod; +ve	-	+	-	-	-	-	-	-	+	*B. cereus*
		White Pin Point	Cocci; +ve	+	-	-	-	+	-	-	-	-	*S. lactis*
		Cream	Cocci; +ve	+	-	-	-	+	-	-	-	-	*S. lactis*
	Bhayzokul	White Round Big	Rod; -ve	-	+	-	+	+	-	+	+	+	*P. vulgaris*
		White Point Point	Cocci; +ve	+		-	-	+	-	-	-	+	*S. lactis*
Chamcham	Muslim Sweets	White	Rod; -ve	+	-	-	-	+	-	+	+	+	*K. pneumoniae*
		Yellow	Rod; +ve		+	-	-	-	+	-	-	+	*B. cerus*
	Bikram pur	Pin Head White	Cocci; +ve	+	-	-	-	+	-	-	-	-	*S. lactis*
		Pin Head Yellow	Cocci; +ve	+	-	-	-	+	-	-	-	-	*S. lactis*
	Bhayzokul	Yellow	Cocci; +ve	+	-	-	-	+	-	-	-	-	*S. lactis*
		Yellow	Rod; -ve	+	-	-	-	+	-	+	+	+	*K. pneumoniae*
		White	Rod; -ve	+	-	-	-	+	-	-	-	+	*E. coli*
Sandesh	Muslim Sweets	Cream	Rod; -ve	+	-	-	-	+	-	+	+	+	*K. pneumoniae*
		White Pin Head	Cocci; +ve	+	-	-	-	+	-	-	-	-	*S. lactis*
	Bikram pur Bhayzokul	White Dispersed	Cocci; +ve	+	-	-	-	+	-	-	-	-	*S. lactis*
		Cream	Cocci; +ve	+	-	-	-	+	-	-	-	-	*S. lactis*
		White Pin Head	Cocci; +ve	+	-	-	-	+	-	-	-	-	*S. lactis*
	Bhayzokul	Dispersed	Rod; -ve		+	-	-	+	-	+	+	+	*K. pneumoniae*
		White Pin Head	Rod; -ve		+	-	+	+	-	+	+	+	*P. vulgaris*
		Orange Round	Cocci; +ve	+		-	-	+	-	-	-	-	*S. lactis*
Kachagolla	Muslim Sweets	Pin Point	Rod; +ve	-		-	-	-	-	-	-	+	*Corynebacterium sp.*
		Pale rigide	Rod; +ve	-	+	-	-	-	-	-	-	+	*Corynebacterium sp.*
		Pin Head	Rod; -ve	+	-	-	-	+	-	+	+	+	*Klebsiella pneumoniae*
	Bikram pur	Round Shape	Rod; -ve	+	-	-	-	+	-	+	+	+	*Klebsiella pneumoniae*
		Pin Head Yellow	Rod; -ve	+	-	-	-	+	-	+	+	+	*Klebsiella pneumoniae*
	Bhayzokul	White Dispersed	Rod, +ve	+	-	-	-	-	-	-	-	+	*Bacilus cerus*
		Orange Big	Rod; +ve	+	-	-	-	-	+	-	-	+	*Bacilus cerus*

Coliform bacteria are one of the major indications of hygienic condition of milk (**8**). Higher coliform bacteria indicate that proper hygienic measures were not usually taken by the above mentioned kachhagolla sample during theKachhagolla preparation. Sen (**9**) reported that the average coliform number in kachhagolla samples of Calcutta and Delhi market were $3.7x10^1$ and $6.2x10^1$ respectively. The average Standard Plate Count for Staphylococcii in urban and rural Sandesh samples was 2.1×10^5cfu/g and 9.46×10^5cfu/g respectively (**10**) reported thatthe product (Sandesh) showed the count 3.0×10^5cfu/g to 7.5×10^7cfu/g. The average coliform in urban and rural Sandesh samples were 2.65×10^2cfu/g and 3.36×10^2cfu/g respectively. The coliform count ranging between 1.0×10^1cfu/g to 1.00×10^2cfu/g in Sandesh samples (**11**). In **Table 1.2.2**, biochemical test was found for the identification for isolates. The presence of *Bacillus* sp. in this study might be due to unhygienic handling. Several dried food sample had been reported to contain some of these organisms (**12-13**). Blackey and Priest (**12**) reported that *Bacillus* spp. is common in soils and vegetation, and has been isolated in several countries from wide variety of routine samples of food. The occurrence of this bacterium in the present study is not surprising because of the way sweets and sweet products are handled in the open market in a dusty environment. Its presence therefore could be due to the contamination from many sources which may include dust, air and water. Pederson (**14**) stated that spores of *bacillus* abounds in air and water and this might be one of the reason for microbes that involved during processing. The presence of *E. coli* in sweets and sweet products is a possible cause of food borne diseases. It is a potential pathogen since several strains of Enteropathogenic *E. coli* have been isolated from milk (raw materials of sweets and sweet products) products suspected to be associated with outbreaks of gastroenteritis and food poisoning in human being. This fact is also supported by Sing and Ranganathan (**15**). The incidence of the presence of other bacterial spp. in this study might be due to the method of sweets and sweet products manufacturing, handling and sale which are entirely based on the traditional system. Such system could provide a favorable environment forbacterial contamination. The unclean hands of workers, poor quality of milk, unhygienic condition of manufacturing unit and water supplied for washing the utensils could be the sources for accelerating thebacterial contamination of milk products beside the post manufacturing contamination. Sweetmeat prepared from milk is an integral part of the culinary habits of people all over India. But these products are extremely vulnerable to contamination with spoilage and pathogenic organisms as well as toxic metabolites of microbial origin if subjected to advertent abuse during their production and processing. More than 200 food borne diseases are now recognized and most of them require specific laboratory diagnosis (**16**). The microbiological safety and quality of foods are directly related to identify the number of microorganism present in the products. Coliforms have probably got more attention than most other groups of bacteria on account of their importance as indicator organisms for predicting unhygienic conditions during production and processing. Among Coliforms, *E. coli* has attracted much attention as a potential pathogen since several strains of Enteropathogenic*E.coli* have been isolated from raw and pasteurized milk and milk products suspected to be associated with outbreak of gastroenteritis and food poisoning in human being (**15**). The technology of Rasmalai making until today is confined in the hands of a few artisans who consider it more of an art than science and do not want to divulge the technology to others. In this way they have maintained their monopoly in the business. Apart from this, it appears that any dairy or food scientists to standardize the technology of Rasmalai production have initiated no work so far. Somehow or the other the product escaped their attention (**17**).

The numbers of viable bacteria of cow milk and buffalo milk were 3.84±0.05 and 3.77±0.07 log cfu/g, respectively and there was a non-significant ($P>0.05$) difference between them. So if the Rasmalai prepared in a good sequence with good quality control practice, then it will be lower in bacterial content (**18**). The number of total viable count per ml of Rossomalai samples of Milk Vita made rossomalai, Bhagabati Pera Bhandar, Comilla, Alauddin Sweetmeat, Dhaka, Islamia Mistanno Bhandar were $4.1x10^3$, $5.5x10^3$, $5.2x10^3$ and $6.0x10^3$, respectively. There was a significant difference ($P< 0.01$) of total viable content of the samples indicates that the number of total viable count in Milk Vita made rossomalai was the lowest and Islamia Mistanno Bhandar was the highest among four samples (**17**). It was found that coliform bacteria were absent in all types of Rasmalai samples. Coliform bacteria are the indicators of hygienic conditions present during the manufacturing of the product. Therefore, coliform count indicates that strict hygienic measures were taken during Rasmalai preparation (**18**). Milk is very nutritious and perhaps an indispensable food for human being. But in this area of industrialization, food habit of common people is changing. They are preferring moor healthy, delicious foods to fresh raw foods. Hence, milk is converted to various products. Yoghurt is one of them locally called Dahi. About 4% of the total milk produced in Bangladesh is used for the preparation of Dahi (**19**). Dahi is prepared by using mixed culture of *Streptococcus lactis, Lactobacillus bulgaricus, Streptococcus thermophilus* and

Streptococcus citrophilus (**4**). Different types of dahi such as sweet dahi, sour dahi and flavoured dahi are usually found in the market. Shakeel et al (**20**) used successfully mango juice to enrich the quality of sweet dahi. Soya milk and sunflower protein could be used in the manufacture of yogurt (**21**). The total viable bacterial count per mL. of different types of Dahi prepared from cow, buffalo and goat milk were 5.878±0.03, 5.996±0.05 and 5.859±0.05 (Log value), respectively.

Statistical analysis showed that significant differences ($p<0.05$) were existed among the different Dahi samples. Average highest total viable count was recorded for Dahi samples of buffalo milk, which that was lowest for goat milk Dahi (**4**).The total viable bacteria contained per mL of Laboratory made sweet Dahi samples were within the range of 6.2 to 6.3 (Log value) (**22**). The average coliform count per ml of cow, buffalo and goat milk Dahi were 1.75±1.20, 2.70±0.21 and 1.00±1.15 (Log value), respectively. Average highest colifom count per mL was found in buffalo milk Dahi, followed by cow milk Dahi and it was lowest for goat milk Dahi. Statistical significant differences were found in respect of coliform count per mL of different Dahi samples (**4**). Islam (**23**) found that coliform count per mL of Laboratory made Dahi was 2.64±2.05 (Log value). The presence of coliform organisms in Dahi samples indicatethat contamination during its production and handling may cause public health problems. The possible sources of contamination of product are uncleaned hands of the manufactures, poor quality water used to clean earthen pots and exposure of the product to open air during setting of curd. The number of coliform per ml of Rossomalai of Milk Vita made rossomalai, Bhagabati Pera Bhandar, Comilla, Alauddin Sweetmeat, Dhaka, Islamia Mistanno Bhandar samples were 1.33±0.47, 3.33±1.24, 2.33±0.47 and 3.00±0.81, respectively. Statistical analysis showed that there was significant not difference ($P< 0.01$) of coliform content within four different samples. Coliform bacteria indicate that proper hygienic conditions were not usually taken during the Rossomalai Preparation (**17**). Adequate consumer protection can be achieved by measuring the microbiological data of end product. To improve the microbiological quality of Sandesh and Kalakand precaution against contamination must be taken at different Critical Control Points (**24**). Coliforms have probably got more attention than most other groups of bacteria on account of their importance as indicator organism for predicting unhygienic conditions during production and processing at the same time some strains of *S. aureus* are gaining entry into milk products and cause acute gastroenteritis via the production of heat stable enterotoxins. Millions of people are the victims of food borne illnesses resulting from toxins produced by *S. aureus* ingested with food.

In present study, it was observed that the mean bacterial counts of these sweet and sweet product samples from all the three brands obviously exceeded the maximum recommended standards by the International Commission on Microbiological Specification of Food (**25**). According to this agency the acceptable limit of mesophilic aerobic bacteria in dried food products should not exceed a maximum of 10^3cfu/g whereas it was found that the counts in all the dried samples taken in present study was considerably high which are unacceptable for human consumption (**26-28**). The generally observed high microbial count in this study could be attributed to the influence of environmental factors on the microbial populations, which have been shown to play a significant role in affecting the quality of food products (**29-34**). The cultural, morphological and biochemical characterization of the microbial isolates recovered from the sweets and sweet products analyzed in thisstudy indicated the presence of *Streptococcus* sp., *Bacilus* sp., *Klebsiella* sp., *Proteus* sp., *E. coli* and *Corynebacterium* sp. Particularly *Streptococcus* sp, *E. coli* and *Bacillus* sp. are important in present study because these are known as causative agents for food poisoning and intoxication (**26**). The presence of these bacteria may be due to the unhygienic environmental conditions and poor processing and handling.

1.2.4CONCLUSIONS AND RECOMMENDATIONS

This investigation was undertaken with a view to assess the microbiological quality of traditional sweets and sweet products. The results revealed that the bacterial count of the sweets and sweet products examined are above the maximum allowable limit for consumption (10^3cfu/g) indicating a public health concern. These high counts are suggestive of heavy bacterial contamination of the sweets during handling since the sweets might have absorbed adequate supply of moisture, which could have contributed tothe development as well as multiplication of these contaminants. From the results of the study, it is evident that due to processing, handling, marketing and unhygienically selling the products to the consumers could expose the consumers to health problems. This should completely be discouraging, so as to produce consistently-consumed sweets and sweet products with high quality and low microbial load. Only sweets and sweet products that are aseptically processed should be sold which could be achieved by:

• Protecting all personnel, the surrounding environment and the packaging materials from contamination with dust and/or other sources of microorganisms during handling according to recommendation of Bangladesh Standard Testing Institute (BSTI).

• Proper washing and sanitization of all equipment and other utensils so as to prevent the spread of diseases (infections) as has been campaigned by the National Institute of Preventive and Social Medicine (NIPSOM).

1.2.5REFERENCES

1. Karthikeyan N and Dhanalakshmi B. 2010. Hygienic Quality of Indian Sweet Milk Products from Different Sources. Bangladesh J. Microbiology, 27(2): 32-37.

2. Sen CD and Rajorhia SG. 1989a. Studies on quality characteristics of Sandesh marketed in Calcutta, Indian J. Dairy Science, 4(1): 241-245.

3. Alam MM, Rahman SMR, Mannan AKMA and Shams-ud-Din M. 2002. Quality Attributes of Kachhagolla- A Delicious Indigenous Milk Product of Bangladesh. Pakistan J. Biological Sci., 5(6): 725-527.

4. Nahar A, Al-Amin M, Alam SMK, Wadud A and Islam MN. 2007. A Comparative Study on the Quality of Dahi (Yoghurt) Prepared from Cow, Goat and Buffalo Milk. International J. Dairy Sci., 2: 260-267.

5. Singh Pand Prakash A. 2008. Isolation of Escherichia coli, Staphylococcus aureus and Listeria monocytogenes from Milk Product Sold under Market Conditions at Agra Region. Acta Agriculturae Slovenica, 92(1): 83–88.

6. Cappuccino JG and Sherman N. 1999. Microbiology- A Laboratory Manual. Wesley publication, 4th Edi., pp 129-183.

7. Boor KJ, Brown DP, Murphy SC, Kozlowski SM and Andlar DK. 1998. Microbiological band chemical band quality of raw milk in New York State. J. Dairy Sci., 81: 1743-1748.

8. Rahman, MS, Islam MS, Hassan MN and Sarker EA. 2000. Quality of Packaged Fluid Milk Available in Retail Market of Mymensingh Town. Bang. J. An. Sci., 29: 163-172.

9. Sen CD. 1992 Studied on kachhagolla a delicious milk sweetmeat of Bengal. Kachhagolla, a type of Sandesh. Indian Dairyman, 44: 249- 251.

10. Sen CD and Rajorhia SG. 1989. Quality characteristics of Sandesh marketed in Calcutta; Indian J. Dairy Science. 42(2): 241 – 242.

11. Misra KA and Kulia KR. 1988. Microbiological quality of Burfi and sandesh. Asian J. Dairy Res., 7(1): 51-55.

12. Blackey LJ and Priest FG. 1980. The occurrence of Bacillus cereus in some dried foods including pulses and cereals. J. App. Bacterio, 4(8): 297-302.

13. Aboloma, RI. 2008. Microbiological analysis of bread samples from bakery to sale points in Ado- Ekiti, Ekiti State, Nigeria. Biol. Environ. Sci. J. Tropics, 5(3): 77-81.

14. Pederson CS. 1979. Microbiology of food fermentation. 2ndedi. AVI publishing company limited, Connecticut. pp 242-249.

15. Singh RS and Ranganathan B. 1978. Incidence and distribution of Escherichia coli in Indian products; Indian J. Dairy Sci. 31: 1.

16. Prasad DN. 1998. Hygenic and safety aspects of dairy products; Indian Dairyman. 50: 12.

17. Islam Z, Rahman SMR, Alam MM, Ali M and Mannan YAKMA. 2003. Manufacture of Rossomalai and Its Quality Attribute: An Indigenous Milk Sweetmeat of Bangladesh. Pakistan J. Nutrition, 2(5): 300-304.

18. Sayedatunnesha, AW, Islam MN and Hossain MA. 2008. Comparative Study of the Quality of Rasomalai Manufactured from Cow and Biffalo Milk. Bang. J. Anim Sci. 37 (1): 57 -62.

19. Mustafa M. 1997. A study on the preparation of fruit Dahi (Yoghurt). M.Sc. Thesis, Bangladesh Agricultural University, Mymensingh, Bangladesh.

20. Shakeel A, Thompkinson DK and Asgar S. 1994. Manufacture of fruit flavoured filled bio-yogurt. Indian Journal of Dairy Science. 47(8): 698-701.

21. Hardi J and Novakovic P. 1994. The feasibility of yoghurt manufacture using cow milk and soya milk blends. Znanost-i-Praksa-u-Poljoprivaredi-i-Prehrambenoj-Tehnologiji. 22(3): 475-490.

22. Adeyl FMM. 1998. Studies on the physical, chemical and microbiological quality of misti dahi of different districts of Bangladesh. M.Sc Thesis, Bangladesh Agricultural University, Mymensingh, Bangladesh.

23. Islam MDS. 1999. A comparative study on the quality of Laboratory prepared and local village market Dahi. MS Thesis, Bangladesh Agricultural University, Mymensingh, Bangladesh.

24. Roy R, Chakrabarti J and Gandhi SR. 1998. Microbiological quality assurance of milk sweets (Sandesh and Kalakand) by HACCP; Indian Dairyman. 50: 4.

25. International commission on Microbiological Specifications of Food (ICMFC) 1978. Microorganisms in Foods, University of Toronto Press, Canada. 1: 110-117.

26. Food and Agricultural Organization (FAO) 1979. Manuals of food quality control 4. FAO Food and Nutrition Paper, United Nations, Rome, Italy. Microbiological Analysis 14(4), A1-F10.

27. Food and Agricultural Organization (FAO) 1993. Codes of Principles concerning milk and milk products. Report of a Joint FAO/WHO Expert Consultations of Microbiological Specifications for Foods, Rome, Italy. FAO (1975-1977) –EC/Microbiol/75/Report and 77/Reports.

28. World Health Organization (WHO) 2003. Microbiological Aspects of food hygiene. Report of a WHO ExpertCommittee with the participation of FAO. WHO Technical Report Series No. 598.

29. Owhe-oreghe UB and Afe OE. 1993. Bacteriological examination of some ready-to- eat foods marketed in Ekpoma, Edo State, Nigeria. Nigerian Food J., 2: 45-52.

30. Abdullahi IO, Umoh VJand Ameh JB. 2005. Microbiological quality and physico chemical properties of 'balangu', a bulk processed meat in Samaru, Zaria, Nigeria. J. Tropical Biosci., 4: 65-68.

31. Shamsuddeen U and Ameh JB. 2008. Survey on the possible critical control points during the production of 'balangu' in Kano. Bayero J. Pure and Appli. Sci. 1(1): 76-79.

32. Shamsuddeen U, Ameh JB and Oyeyi TI. 2008. Survey on the possible critical control points during the production of 'dambun nama' in Kano. Biol. Environ. Sci. J. Tropics, 5(4): 1-5.

33. Oyeyi TI and Lum-nwi MEF. 2008. Bacteriological quality of some street-vended foods in Bayero University campuses, Kano, Nigeria. Biol. Environ. Sci. J. Tropics, 5(4): 239- 243.

34. Wada-kura A, Maxwell RG, Sadiq HY, Tijjani MB, Abdullahi IO, Aliyu MS and Adetunji OA. 2009. Microbiological quality of some ready-to-eat foods and fomites in some cafeterias in Ahmadu Bello University, Zaria. Biol. Environ. Sci. J. Tropics, 6(1): 6-9.

1.3 IRRADIATION EFFECT ON RITA FISH KEBAB PRESERVATION

SOURCE: *Md. Harunur Rashid, Zakaria Ahmed, Harun—Or—Rashid, A. N. M. Fakhruddin and Naiyyum Choudhury. 1998. Combination effect of irradiation and low temperature on the survival of microbes in Rita fish kebab. J. Asiat. Soc. Bangladesh, Sci. 24(2): 309-318.*

1.3.1 INTRODUCTION

Rapid growth in urbanization is making fast food or convenience food very popular among the working people and fish, meat and vegetable-based convenience foods are now sold in many places of Dhaka city and its surrounding areas. The preparation of such food products involve prior processing of raw materials with different ingredients. Preparation, processing and preservation of different types of these convenience foods have of late received much attention because of theirever-increasing demand. This hasalso necessitated the measure of precaution in terms ofmicrobiological safety of the products. Preservation and shelf lifeextension of fast food like fish kebab largely depends upon source of the fish. microbial population, nature of handling and processing. Depending on the habitat and other environmental factors, a wide range of variation in distribution of microflora in fish products has been reported (**1-2**). Cann (**3**) reported that primary contamination occurred upto the limit of harvesting whereas secondary contamination began at the start of processing as a result of contact with workers" and processing equipment. Fish product like fish kebab is also important. in this regard as ready-to-eat convenience food which is subject to preservation by combination treatment with gamma irradiation and low temperature storage. As such,combination treatment involving irradiation, has gained popularity as preservation technology. Irradiation of prepackaged products has the advantage of penetrating through the product and as such, any mishandling, which may result in the escape of microbes can be eliminated. The present investigation was mainly directed to minimize microbial load and to extend the shelf life of fish kebab by gamma irradiation in combination with low temperature storage.

1.3.2MATERIALS AND METHODS

Preparation of fish kebab of Rita fish (*Rita rita*) was conducted in laboratory scale where flour 7.5g, ginger paste 218g, Cumin powder 1.5g, Clove powder 0.5g and table salt 0.5g were well mixed with the blanched fish flesh and finally 4-5cmx3.5cmx1.5cm sized kebabs were prepared. The whole preparation was carried out inhygienic condition. The kebabs were then fried for 5min. in boiled soybean oil and cooled in laminar air flow cabinet.The fried kebabs were then aseptically packed in pre-sterilized (3.0Mrad) polythene pouches. In each pouch 12 kebabs were separately sealed. A gamma irradiation source of 50,000 $CiCo^{-60}$ at the institute of Food and Radiation Biology (IFRB) of Atomic Energy ResearchEstablishment, Savar, Dhaka was used. Two sets of kebab packetseach were then irradiated at a dose of 2.5kGy and 5kGy. Whiletwo sets were kept as control (un-irradiated]. All the control and un-irradiated sample packets were then stored at ambient and 4°C temperature in freezer. The samples were examined for microbiological status and organoleptic quality before and after being kept for preservation.For microbiological analysis. 10g of samples (control and irradiated) of fried fish kebabs were taken aseptically in sterile homogenizer with 90 ml sterile normal saline (0.9% NaCl) and homogenized for 3min. Making 10^{-1} dilution and by using tenfold serial dilution technique (**4**) upto 10^{-1}dilutions were made. Standard spread plate count technique (Cappuccino and Sherman-1986) was followed throughout this study for the determination of microbial count of control and irradiated fish kebab samples at ambient and 4°C temperature storage. Nutrient agar (NA), McConkey agar (MA), Staphylococcal Agar (SA), Starch Ampicillin Agar (SAA), Membrane Fecal Coliform Agar (MFCA), Total Viable Coliform Count (TVCC), Total Viable Staphylococcal Count (TVSC), Total Viable Aeromonas Count (TVAC),Total Viable Membrane Fecal Coliform Count (TVMFCC) and Total Viable Fungus Count (TVFC), respectiifely. All the media were autoclaved with due respect and pH was adjusted (Cappuccino and Sherman 1986). The NA, MA, SAA, and SA media were incubated at 37°C for 24 to 72h whereas PDA media was incubated at 28°C for 24h (Cappuccino and Sherman 1986).In all the cases, the microbial count was expressed in colony forming unit per gram (cfu/g). The water activity (a_w) of the samples was estimated at different time intervals. For estimation of moisture content in the samples, porcelain made crucible was heated at 105°C for 1h and then kept in a desiccator until the crucible reached the room temperature, then 10g of sample was taken in the crucible and dried at 105°C for 3h. Then the crucible was kept in a desiccator until it reached 'the room temperature and weight of the sample was taken. The process was continued until constant weight was obtained.

$$\text{Percentage of moisture} = \frac{\text{Weight loss of the sample}}{\text{Original weight of the sample}} \times 100$$

Sensory evaluation for the determination of shelf lifeof kebab was done by organoleptic test scored on a 9-point hedonic scale as described by Patterson and Stevenson (**5**). The samples were evaluated for taste, odor, color and texture. Detailed organoleptic tests were performed by six trained judges of concerned area of different days of storage.The hedonic scales were as follows:

Point	Remarks	Point	Remarks	Point	Remarks
1	Dislike	4	Dislike slightly	7	Like moderately
2	Dislike very much	5	Neither like or Dislike	8	Like very much
3	Dislike moderately	6	Like slightly	9	Like extremely

1.3.3RESULTS AND DISCUSSION

A six-month period of survey was conducted for the determination of radiation and low temperature storage effect on microbes of Rita fish kebab. The initial total viable microbial count, before frying were as follows- bacteria ($1.2X10^{5}$cfu/g); fungus ($3.0X10^{2}$cfu/g); coliform ($3.5X10^{3}$cfu/g); *Staphylococcus* ($1.9X10^{4}$cfu/g), *Aeromonas* ($1.7X10^{3}$cfu/g) and fecal coliform ($1.8X10^{2}$cfu/g). After frying, initial Staphylococcal count was $4.2X10^{2}$cfu/gwhich found to be $1.0X10^{6}$cfu/g and $9.8X110^{3}$cfu/g after 21 days of storage at ambient and 4°C temperature (**Fig. 1.3.1**) respectively. In 2.5 kGy treated sample, the count was decreased to $6.1X10^{1}$cfu/g and the count was $9.1X10^{3}$cfu/g and $9.8X10^{3}$cfu/g, respectively when this treated kebab was stored for 60days at ambient and 4°C temperature respectively. When the fish kebab was subjected to irradiation with 5.0kGy no Staphylococcal growth was observed and contaminated beyond detectable level during storage at both ambient and 4°C temperature. The TVBC of fried fish kebab was initially $4.6X10^{3}$cfu/g. Due to irradiation treatment with 2.5 and 5.0kGy, the TVBC decreased to $7.5X10^{1}$cfu/g and less than 10, respectively (**Fig 1.3.2**). When these kebabs,both control and irradiated (2.5 81 5.0kGy), were subjected tostorage at ambient temperature, it was observed that the count was gradually increased and after 60 days of storage, the controlsample showed the TVBC of $5.2X10^{3}$cfu/g whereas 2.5kGy and 5.0kGy treated samples showed 1.0X106cfu/g and $1.9X10^{4}$cfu/g respectively. During preservation of fish kebab at 4°C the count was increased but at slower rate than ambient temperature storage. After 60 days of storage at 4°C, the TVBC of control sample was observed $9.8x10^{3}$cfu/g and $1.2X10^{3}$cfu/g in 2.5 & 5.0kGy treated samples, respectively (**Fig 1.3.2**).

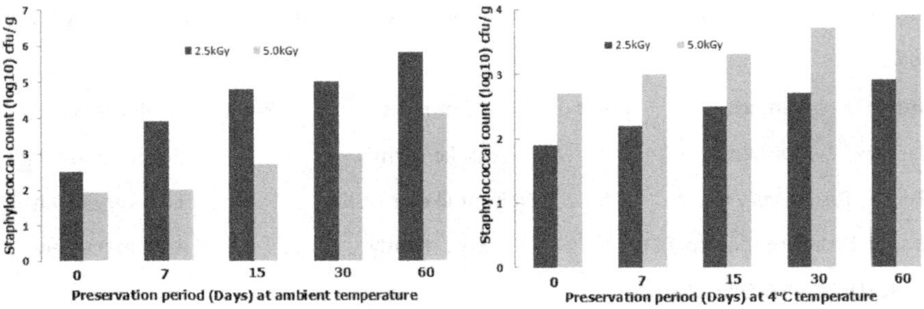

Figure 1.3.1: Effect of storage on total Staphylococcal count of fish kebab irradiated at 2.5 and 5 kGy during preservation at ambient temperature and 4°C. [Md. Harunur Rashid et al 1998]

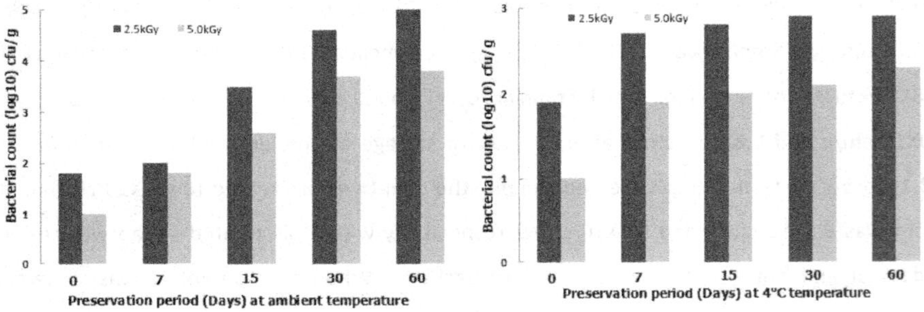

Figure 1.3.2: Effect of storage on total viable bacterial count of fish kebab irradiated at 2.5 and 5.0kGy during preservation at ambient temperature and 4°C. [Md. Harunur Rashid et al 1998]

Initial fungal and coliform count of fish kebab were $5.0x10^1$cfu/g and $1.2x10^2$cfu/g respectively and after 60days of storage at ambient and 4°C temperature, the counts were $7.85x10^1$cfu/g for fungal and $9.8x10^5$cfu/lg and $8.2x10^3$cfu/g for coliform, respectively. Due to irradiation treatment with 2.5 kGy and 5.0kGy no fungal and coliform growth were observed both at ambient and 4°C temperature storage. In fish kebab, the initial *Aeromonas* count was $2.3x10^2$cfu/g, which increased to $1.8x10^5$cfu/g and $4.7x10^3$cfu/g after 60days of storage at ambient and 4°C temperature respectively. Butno *Aeromonas* count was observed in the irradiation treated samples at any temperature after storage period. During the present study, no fecal coliform was found in fish kebab.The initial moisture content of fish kebab was 40.91% andafter 60 days of storage at ambient temperature and 4°C temperature; it was decreased to 37% and 37.5%, respectively. The water activity (a_w) in fish kebab was also decreased from 0.941 to 0.932 at ambient and 0.929 at 4°Ctemperature (**Table 1.3.1**). Sensory evaluation for taste, odor, color and texture of fish kebab were investigated during storage period on the basis of Hedonic score (**Fig1.3.3 and 1.3.4**) where the sensory properties of irradiated fish kebab preserved at 4°C shows superior quality than that of others.

Table 1.3.1 Estimation of moisture content and water activity of fish kebab during storage at ambient and 4°C temperature in different days. [Md. Harunur Rashid et al 1998]

Storage	**At ambient temperature**		**At 4°C temperature**	
	Moisture content (%)	Water activity (a_w)	Moisture content (%)	Water activity (a_w)
0	40.91	0.941 at 24.8°C	40.91	0.941 at 24.8°C
7	39.30	0.936 at 27.0°C	39.60	0.933 at 25.0°C
15	38.00	0.934 at 27.0°C	39.00	0.931 at 25.6°C
21	37.00	0.932 at 26.4°C	37.50	0.901 at 26.0°C

In 1993, Choudhury et al (**6**) reported thatthe total microbialload in fish kebab collected from (ordinary shop was higher (10^4-10^6cfu/g) in comparison to that of sophisticated shop (10^3-10^4cfu/g), whereas total coliform countcollected from ordinary and sophisticated shops were beyond detection limit. Our present study also supported the result of Loharanu (**7**) who reported that the combination of various conventional treatments with irradiation is being successfully applied for preserving certain food products. In addition, the combination treatments act in a way of exerting additive or synergisticeffect on spoilage microbes. sensitive spores to irradiation orinhibit the outgrowth of any surviving spores after the treatment.It was reported by Choudhury et al (**6**) that kebabs prepared at laboratory scale according to selected formulation and irradiated at a dose 5kGy could be stored for 3 months at ambient temperature. In the present investigation, it was thus observed that there was 1 to 3 log reduction. of total bacterial count of fried kebab when treated with 2.5kGy and 5.0kGy dose of irradiation. In 3 weeks' time the bacterial load rose to 10^9cfu/g for untreated kebabs stored at ambient temperature, whereas the 2.5 and 5.0kGytreated stored under similar condition showed 10^6 and 10^4cfu/g which was within the borderline of acceptance. When the radiation treated kebabs were stored at 4°C the microbiological quality was far superior to that observed for ambient temperature stored kebabs (**Fig. 1.3.1**). The microbiological load of irradiated kebabs was 10^2-10^3cfu/g even after 60 days of storage. However, the untreated kebabs were found to have higher microbial load (10^5cfu/g) within 15 days of storage at 4°C temperature. No fungus could be detectedin the irradiated kebabs whereas fungal" count rose to about 10^3cfu/g in 60 days for ambient temperature storage and to about 10^2cfu/g in 4°C temperature storage. The coliform count of ambient and 4°C stored untreated kebab rose to around 10^5cfu/g and 10^3cfu/g respectively, whereas no coliform could be detected in the irradiated kebabs. Similarly, Staphylococcal and *Aeromonas* count of the irradiated kebabs were low compared to that of irradiated kebabs.

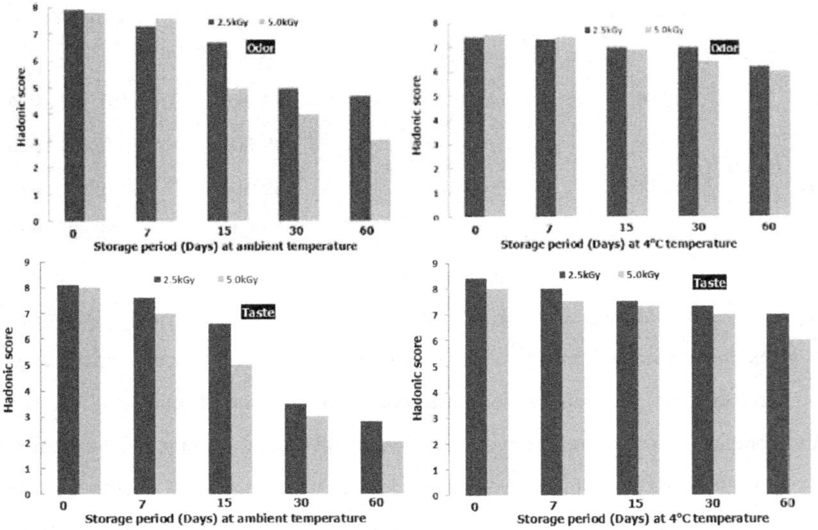

Figure 1.3.3: Sensory evaluation (taste and odor) of untreated and irradiated fish kebab during storage at ambient temperature and 4°C. [Md. Harunur Rashid et al 1998]

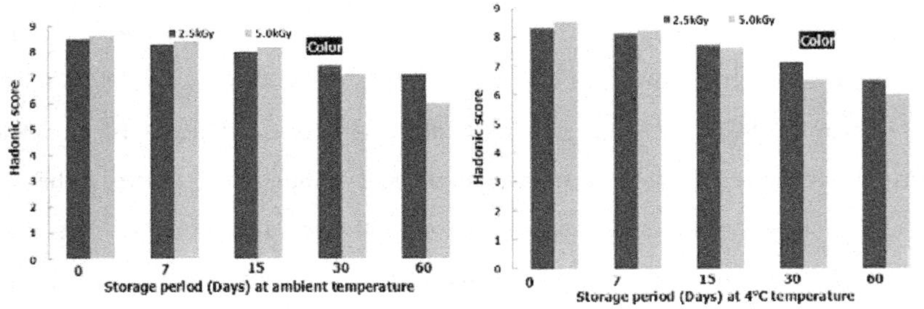

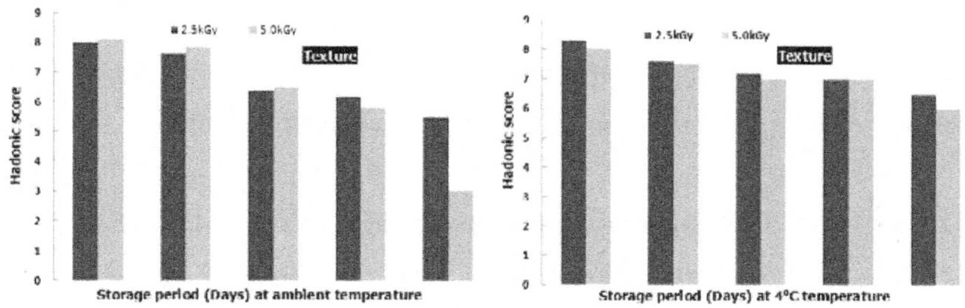

Figure 1.3.4: Sensory evaluation (color and texture) of untreated and irradiated fish kebab during storage at ambient temperature and 4°C.

Sensory evaluation of irradiated and untreated kebabs showed significant improvement in taste, odor, color and texture of irradiated fried kebab over that of untreated samples. Kebab treated with 5.0kGy dose were acceptable upto 15days if stored atambient temperature and upto 60days if stored at 4°C on the taste, odor, color, and texture score of Hedonic scale. Thus, it is possible to preserve semi cooked kebabs upto 2weeks at ambient temperature if it is irradiated at 5.0kGy and for similar irradiation treatment, the prepared kebabs can be stored for 2months at 4°C. It was observed from the present study that combination of irradiation (5.0kGy) and low temperature (4°C) can greatly enhance the shelf life of fish kebabs. However, an element of risk isassociated with such processing. Since the prepared kebabs are packed in polythene packages and then sealed, the environment inside the package is almost anaerobic. This situation increases the risk of anaerobic growth of spore former (*Bacillus, Clostridium* etc.) surviving the processing and irradiation treatment. This can be avoided by controlling water activity of kebabs. It was found in the present study that water activity of kebab stored (4°C) for 30days was 0.901 (**Table1.3.1**) which is quite safe for human consumption as spore former cannot grow at water activity below 0.92 (**1**). The present work indicates the possibility of short and longterm preservation of ready-to-eat convenience food. Even one-week extension of shelf life of such product at ambient temperature opens the possibility of distribution of such microbiologically safeproducts to remote places for consumption by working people of such food avoiding the time and energy in traditional cooking process.

1.3.4 REFERENCES

1. Shewan JM. 1976. The bacterial action. In: Proceedings of Tropical Institute Conference on handling, processing, and marketing of tropicalfish. Tropical Product Institute, London. pp. 51-66.

2. Hess E. 1934. Contribution of Canadian Biology and Fisheries. 8: 461-474.

3. Cann DC. 1977. Bacteriology of shelf fish with references to international trade. In Handling, Processing and Marketing of Tropical Fish Tropical Product Institute, London, pp. 46-48.

4. Cappuccino JO and Sherman N. 1936. Microbiology: A Laboratory Manual. The Bgnjamm/Cummings Publishing Company, Inc. pp. 125-179. 445-451.

5. Patterson MF and Stevenson MIl. 1992. Effect of gamma irradiation on the Microbiological and organoleptic quality of readymade chilled meals. J. Food Protection 49:189.

6. Choudhury N, Siddiqui AK, Chowdhury NA, Youssouf QM. Rashid H. Begum AA and Alam MK. 1993. Preservation of semi perishable foods and development of convenience food by combination of irradiation and other physico-chemical treatments. Report presented at the second FAO/LAEA research coordination 'meeting (RCM) on irradiation in combination with other process for improving food quality. Food Research and Development Center, St./Hyacinth, Quebec. Canada. 28th June and July, 1993.

7. Loharuanu. P. 1973. Radiation preservation of food. IAEA-SM-166/67. Vienna. pp.4-27.

CHAPTER TWO

2.1 MEDICINAL VALUE AND ANTIMICROBIAL ACTIVITY OF SOME SELECTED PLANT PRODUCTS

2.2 MEDICINAL VALUE AND ANTIMICROBIAL ACTIVITY OF SOME SELECTEDPLANT

2.1 MEDICINAL VALUE AND ANTIMICROBIAL ACTIVITY OF SOME SELECTED PLANT PRODUCTS

SOURCE:

1. *Ummey Nahor and Zakaria Ahmed. 2012. Antimicrobial Activity of Phyllanthus Emblica and Allium Sativum: Comparative Analysis of Antimicrobial Action of Crude and Ethanolic Extract of These Natural Plant Products. IOSR Journal of Pharmacy and Biological Sciences (IOSR-JPBS). 4(3): 21-26.*
2. *Zakaria Ahmed and Ummey Nahor. 2012. Benificial Uses and Antimicrobial Activity of Phyllanthus Emblica, Achyranthes Aspera and Allium Sativum- A Mini Review. IOSR Journal of Pharmacy and Biological Sciences (IOSR-JPBS). 3(4): 28-32.*

2.1.1 INTRODUCTION

Bioactive compounds from diverse sources have been isolated and characterized around the world. Systematic screening of plant materials represents an important effort to find new bioactive compounds with the needed therapeutic potential to fight against pathogenic microorganisms particularly with respect to those that are hospital based. Nature has been a source of medicinal agents for thousands of years and an impressive number of modern drugs have been isolated from natural sources. During the last two decades there has been an up search in search for new plant derived drugs containing the medically useful alkaloids, glycosides, polyphenolics, steroids and terpenoids derivatives. Plants used for traditional medicine contain a wide range of substances that can be used to treat chronic as well as infectious diseases. Clinical microbiologists have great interest in screening of medicinal plants for antimicrobial activities and phytochemicals as potential new therapeutics. Researchers identified 119 secondary metabolites isolated from the plant that are being used globally as drugs. It has been estimated that 80% of the world's population still use the traditional medicines for their primary health care needs. Many important modern plant drugs such as the winblastin and wincristiene have been discovered by following from traditional medicines (**Raghu and Ravindra 2010**). In the indigenous cultures higher plants have formed the basis for the treatment of the diseases since the earliest time. Accuracy in recording or observing the medical use of a plant determining whether the ethanomedicinal use can be demonstrated under the specific condition in the laboratory (**Ghosh et al 2008**). Chemical characterization of the compounds and the role of the placebo effect are important issues that need to be verified in the development of drugs ofplant origin (**Raghu and Ravindra 2010**). The one of the largest group of chemical produced by plants are the alkaloids and their amazing effect on humans has led to the development of powerful pain killer medications (**Egwaikhinde et al 2008, Raffauf 1996**). The dried fruits have a good effect on hair hygiene and have long been respected as an ingredient of shampoo and hair oil (**Thakur et al 1989**). *Acalypha indica* (Euphorbiaceae), *Aerva lanata, Jussex schult*(Amaranthaceae), *Phyllanthus amarus* (Euphorbiaceae), *Phyllanthus emblica* (Euphorbiaceae), *Cassia auriculata* (Caesalpiniaceae) and *Caesalpinia pulcherrima* Sw (Caesalpiniaceae) have high medicinal properties and active compounds (**Pranithanchai et al 2009, Rao et al 2005, Schiebinger 2004**). The antibacterial studies of the above medicinal plants were already investigated against some common human pathogenic bacteria (**Anushia et al 2009**). Population rise, inadequate supply of drugs, prohibitive cost of treatments, side effects of several allopathic drugs and development of resistance to currently used drugs for infectious diseases have led to increased emphasis on the use of plant materials as a source of medicines for a wide variety of human ailments. Global estimates indicate that 80% of about 4 billion populations cannot afford the products of the Western Pharmaceutical Industry and have to rely upon the use of traditional medicines, which are mainly derivedfrom plant material (**Brindha et al 2008**). Practitioners of traditional medicine believe that the constituents of plants are unique as they contain both active ingredients and "non-active" components that play a role in enhancing the well-being of their patients (**Ureghe et al 2010**).

In view of increasing resistance to existing antimicrobial agents, some research has been performed worldwide to identify herbal drugs, as they are very important sources for discovering some new agents for treating various ailments related to bacterial infections (**Pan et al 2012**). With increasing resistance, antibiotics are occasionally associated with adverse side effects to the host, including hypersensitivity, immune-suppression and allergic reactions (**Graul et al 2009**). These adverse reactions lead to discovering some natural antimicrobial agents which will be effective against pathogenic microorganisms with minimal side effects. One possible strategy is the rational localization of bioactive products from folk medicines, with the hope that systematic screening of these will result in the discovery of novel moiety with potent and useful activities against microbes. With this respect, there is an increasing demand for plant-based therapeutics in both developing and developed countries due to a growing recognition that they are natural products, non-narcotic and, in most cases, easily available at affordable prices with no side effects (**Winslow and Kroll 1998**). Plants are rich in a wide variety of secondary metabolites, such as tannins, terpenoids, alkaloids, and flavonoids, which have been found in vitro to have antimicrobial properties (**Cowan 1999**) and have an almost limitless ability to synthesize aromatic substances, most of which are phenols or their oxygen-substituted derivatives (Pyrrole pigments, isoprenoid compounds and phenolic plant constituents) (**Geissman 1963**). Many of the plants could be used as stimulants, hallucinogens or as medicine because of the presence of unique or rich biological-active plant chemicals. Chemicals that make a plant valuable as medicinal plant are: alkaloids (compounds has addictive or pain killing or poisonous effect and sometimes help in important cures), glycosides (use as heart stimulant or drastic purgative or better sexual health), tannins (used for gastro-intestinal problems like diarrhea, dysentery, ulcer and for wounds and skin diseases), volatile/essential oils (enhance appetite and facilitate digestion or use as antiseptic/insecticide and insect repellant properties), fixed oils (present in seeds and fruits could diminish gastric/acidity), gum-resins and mucilage (possess analgesicproperty that suppress inflammation and protect affected tissues against further injury and cause mild purgative), and vitamins and minerals (fruits and vegetables are the sources of vitamins and minerals and these are used popularly in herbals) (**Ghani1998**). Triphala ia a mixture of Bibhitaki (*Terminalia bellirica*) with Haritaki (*Terminalia chebula*) and Amla (*Phyllanthus emblica*) which is rejuvenating in nature (the Ayurvedic Formulary of India 2003). Bibhitaki is very useful for hair, throat, and eyes related problems. It is known to boost the immune system of the body and it also helps in maintaining the balance of "Tridosha (humor)" especially Kapha (Water). Bibhitaki, the wonder medici ne has several health benefits, e.g.- prevents hair loss and removes dandruff, helpful in constipation, removes acne and ulcers, fights with skin allergies, prevents cough and cold, boosts immunity, prevents diabetes and help in weight loss. Haritaki fruits are highly nutritious for human health as they contain various vitamins, minerals and proteins. They are an excellent source of vitamin C. These fruits are also rich in several minerals including selenium, potassium, manganese, iron and copper. Henna isused as a topical treatment for fungal diseases of the skin, particularly those caused by Tinea (**Fariba et al 2010**). It is also used to relieve rheumatic pain. A review of experimental animal studies indicates that henna extract enhances wound contraction, reduces epithelialization time, skin cancer chemo preventive activity (**Pradhan et al 2012, Kapadia et al 2013**); it significantly increases the weight of granulation tissue and, thus, its use has been suggested for improving wound healing (**Nayak et al2007**). This study was based for the scientific evaluation and use of selected natural products or plant materials such as Trifala mixtures of Amla (*Emblica officinalis*), Haritaki (*Terminalia chebula*) and Bibitaki (*Terminalia belerica*), Java plum (*Syzygium cumin*), Cinnamon (*Cinnamomum verum*), Henna (*Lawsonia inermis*), Cumin (*Cuminum cyminum*), which includes study the antimicrobial effect of some natural products (herbal and spice extracts); comparative analysis of antimicrobial activitybetween aqueous and ethanol extract of test products; and determination of MIC and MBC of these extracts.It was expected that since establishment of project, there might have been some impact on its beneficiaries in respect of their socioeconomic issues.

2.1.1.1AMLOKI (*Phyllanthus emblica*)The plant genus phyllanthus (Euphorbiaceae) is widely distributed in most of tropical and subtropical countries.*P.emblica* has been used for the anti-inflammatory and anti-pyretic treatments by the rural population and has been used for the treatment of several disorders such as the Scurvy, Cancer and Heart diseases. The important constituent of plant leaves hasthe anti-neutrophilic activity and anti-platelet properties in vitro. The extracts also possess several pharmacological properties like anti-viral (HIV, AIDS, Herpes Virus, CMV) antimutagenic, anti-allergic, anti-bacterial activities (**Raghu and Ravindra 2010, Khopde et al 2000**).*P. emblica* contains different class of secondary metabolites (**Calixto et al 1998**). The phytochemical analysis of *P. emblica* reveals the presence of alkaloids, tannins and saponins. These compounds are known to be biologically active. Tannins have been found to form irreversible complexes with proline-rich proteins (**Raghu and Ravindra 2010**) resulting in the inhibition of the cell protein synthesis. Apart from antimicrobial activity exhibited by tannins, they also react with proteins to provide the typical tanning effect. Medicinally, this is important for the treatment of inflamed or ulcerated tissues (**Raghu and Ravindra 2010**). Tannins have important roles such as stable and potent antioxidants. Herbs that have tannins as their main component are astringent in nature and are used for treating intestinal disorders such as diarrhea and dysentery (**Dharmananda 2003**). Medical studies conducted on*P. emblica*fruit suggest that it has antiviral properties (**Udupa 1985**) and also functions as an antibacterial and anti-fungal agent (**Treadway 1994**). The use of amla as an antioxidant has been examined by a number of authors (**Bhattacharya et al 1999, Chaudhuri 2004**). Experiments conducted at the Niwa Institute of immunology in Japan had shown *P. emblica* to be a potent scavenger of free radicals where it was found that *P. emblica* preparations contained high levels of the free-radical scavenger, superoxide dismutase (SOD), in the experimental subjects (**Treadway 1994**). Emblica was shown to reduce UV-induced erythema and had excellent free-radical quenching ability, chelating ability to iron and copper as well as MMP-1 and MMP-3 inhibitory activity (**Chaudhuri 2004**). The juice of the bark of *P. emblica* combined with honey and turmuric is a remedy for gonorrhoea (**Nadkarni and Nadkarni 1999**). The expressed juice of *P. emblica* along with other ingredients is used to cure fits and insanity (**Jayaweera 1980**). In Indonesia, the pulp of this fruit is smeared on the head to dispel headache and dizziness caused by excessive heat (**Perry 1980**). On the other hand, it can be used for anointing and "cooling" the head by mixing with buttermilk (**Treadway 1994**). *P. emblica* has been used for anti-inflammatory and antipyretic treatments by rural populations in its growing areas. *P. emblica* powder is mixed with red sandalwood (*Pterocarpus santalinum*) and prepared in honey to relieve nausea and vomiting (**Treadway 1994**). Tang et al (**2009**) investigated the antimicrobial spectrum and activities of 70% ethanol extract of *P. emblica* fruits and the results indicated that the extract shows little antimicrobial activity against tested mold, but it has certain antimicrobial activity against*Saccharomyces cerevisiae* with a MIC of 8.0%, and presents strong antimicrobial activities against *Bacillus stearothermophilus*, *Bacillus proteus*, *Staphylococcus aureus*, *E. coli* and *Bacillus subtilis* in turn with MICs of 2.0%, 2.0%, 2.0%, 4.0% and 4.0%, respectively.*P. emblica* can be used medicinally for the treatment of diarrhea. Raghu and Ravindra (**2010**) stated that *P. emblica* fruits have been used for various disorders. According to their findings, *P. emblica* methanolic extract exhibited a significant antimicrobial activity where the MIC exhibited against the Gram positive and negative microbes ranges between 0.261 to 0.342. The expressed juice of the fruit along with other ingredients is used to cure cough, hiccough, asthma and other diseases (**Jayaweera 1980**). Chaudhuri (**2004**) reported the effectiveness of a standardized antioxidant fraction of *P. emblica* as a skin lightener. *P. emblica* is also used as a febrifuge, as an anti-inflammatory, anti-diuretic and preventive tonic against loss of hair (Tsarong 1994). Xiaoli et al (**2009**) first reported on the in vitro antimicrobial and antioxidant properties of Supercritical Fluid Extraction (SFE) and methanolic extracts prepared from*P. emblica* where they introduced a unique natural source that possesses strong antimicrobial and antioxidant substances.The SFE extract had a strong and broad spectrum of antimicrobial activity and in some cases, even found to have similar activity to the antibiotics ampicillin and nystatin. On the other hand, the methanolic extract exhibited quite strong antioxidant activities, which were found to be similar, and in some cases even higher than the reference compound butylated hydroxytoluene (BHT).

2.1.1.2 BELY ASRA (*Achyranthes aspera*)*Achyranthes aspera* belongs to the family-Amaranthaceae. It is an annual, stiff erect herb, and found commonly as a weed and used by traditional healers for the treatment of fever, dysentery and diabetes (**Girach and Khan 1992, Liersch 1992**). Leaf decoction for cardiovascular toxicity has been reported (**Han and Un 2003**), and the ethanol crude extract showed high larvicidal activity on the tick larvae against*Boophilis microplus*(**Chungsamarnyart et al 1991**). The root extract is well reputed for its pronounced insect molting hormonal activity and the ethanolic extract of the leaves and stem of the plant inhibited the growth of *Bacillus subtilis* and*Staphylococcus aureus* bacterial strains (**Valsaraj et al 1997**). Roots are used as astringents to wounds, in abdominal tumor and stomach pain (**Ghani 2003**). The benzene extract of the stem bark shows abortifacient activity in the rat (**Bhattarai 1994**). Leaf extracts were reported to possess thyroid stimulating, antiperoxidative and antifungal activity properties (**Tahiliani 2000, Elumalaii et al 2009**). The aqueous and methyl alcohol extracts of the plant also decreased blood glucose levels in normal and alloxan diabetic rabbits (**Akhtar and Iqbal 1991**). It is reported to contain alkaloids, flavonoids, saponins, steroids and terpenoids. The water-soluble alkaloid achyranthine isolated from *Achyranthes aspera* possess anti-inflammatory activity (**Sutar et al 2011**). *A. aspera* is a perennial herb growing upto three meters. Juice of *A. aspera* root is used in the treatment of diabetes; juice from leaf is used to treat earache (**Chopra et al 1986**). Several investigators have reported that the methanolic extracts of leaves *A. aspera* has significant antimicrobial activity against the Gram-Positive (*S. aureus, B. subtilis*), Gram-negative bacteria (*K. pneumoniae, E. coli*) and fungal species (*A. niger, C. albicans*) (**Ramesh et al 2011**). It is also used in the treatment of gonorrhea. The whole part of the plant is claimed to posse's medicinal properties in the traditional medicinal system. Extracts from this plant is given to cure diabetics in mammals. Phytochemical or biological works were scanty in *A. aspera* (**Torssel 1993**). There are scanty reports on its antibacterial activities of this plant. In order to demonstrate the antibacterial efficacy, test was conducted against human pathogenic bacteria including those responsible for causing inflammation. The plant is used in indigenous system of medicine as emenagogue, antiarthritic, antifertility, laxative, ecbolic, abentifacient, anti-helminthic, aphrodisiac, antiviral, anti-plasmodic, antihypertensive, anti-coagulant, diuretic and anti-tumor (**Anonymous 1985**). It is also useful to treat cough, renal dropsy, fistula, scrofula, skin rash, nasal, infection, chronic malaria, impotence, fever, asthma, piles and snake bites (Selvanayagam et al. 1994). The root is astringent, diuretic and antispasmodic. It is used in the treatment of dropsy, rheumatism, stomach problems, cholera, skin diseases and rabies (**Manandhar 2002**). The juice extracted from the root of this plant, mixed along with the root extracts of *Urena lobata* and the bark of *Psidium guajava*, are used in the treatment ofdiarrhea and dysentery (**Manandhar 2002**). The plant is astringent, digestive, diuretic, laxative, purgative and stomachic. The juice of the plant is used in the treatment of boils, diarrhea, dysentery, haemorrhoids, rheumatic pains, itches and skin eruptions.

The ash from the burnt plant, often mixed with mustard oil and a pinch of salt, and is used as a tooth powder for cleaning teeth. It is believedto relieve pyorrhea andtoothache. The leaf is emetic and a decoction is used in the treatment of diarrhea and dysentery. A paste of the leaves is applied in the treatment of rabies, nervous disorders, hysteria, insect and snake bites. Mohinder et al (**2005**) stated that Chloroform and methanol root and shoot extracts of *A. aspera* showed good amount of antibacterial activity against *Klebsiella* sp. Elumalai et al (**2009**) reported that the ethanol and methanol extracts of the leaves of *A. aspera* revealed a significant scope to develop a novel broad spectrum of antifungal herbal formulation.

2.1.1.3 GARLIC (*Allium sativum*)Garlic (*Allium sativum*) is one of those plants that were seriously investigated over the years. It has been used for centuries to fight infections (Onyeagba et al 2006). The early Egyptians used it to treat diarrhea; the ancient Greeks used it to treat intestinal and extra-intestinal diseases,while the ancient Japanese and Chinese used it to treat headache, flu, sore throat and fever. In Africa, particularly in Nigeria, it is used to treat abdominal discomfort, diarrhea, otitis media and respiratory tract infections (**Ankri and Mirelman 1999, Jaber and Al-Mossawi 2007**). Jehan et al (**2011**) stated that water extracted samples of garlic indicated a good range of inhibitory effect against *Salmonella typhi* and butanol extracted sample showed highest activity against *Erwinia carotovora.* Garlic has been shown to inhibit the growth of a variety of microorganisms, not only bacteria but also fungi and viruses. The antimicrobial activity of garlic is believed to be due to the effect of allicin, the main ingredient in garlic, generated by the phosphopyridoxal enzyme allinase (**Susumu et al 1987**). In investigation of the activity against other fungal and bacterial strains by Susumu et al (**1987**), ajoene was found strongly inhibited the growth of*Candida glabrata, C. tropicalis, Trichophyton mentagrophytes, Tricosporon beigelii* and *Saccharomyces cerevisiae*. Deresse (**2011**) reported that *A. sativum* has antimicrobial properties against *S. aureus*. It has both a bacteristatic and bactericidal activity when tested in vitro using crude preparation of garlic. Therefore, garlic may be used successfully for treating food poisoning causative agent like *S. aureus*. A compound, found exclusively in garlic, may control blood sugar levels just as well as insulin but without the need for daily injections (**Hiromu et al 2010**). Garlic has also proposed to treat asthma, candidiasis, colds and antibacterial effect against food borne pathogens like *Salmonella, Shigella* and *Staphylococcus aureus*(**Teferi and Hahn 2002**). Therapeutic use of garlic has been recognized as a potential medicinal value for thousands of years to different microorganisms, such as antifungal, antiviral, antibacterial antihelmantic, antiseptic and anti-inflammatory properties of garlic are well documented. Moreover, garlic extracts exhibited activity against both Gram negative (*E. coli, Salmonella* spp.,*Citrobacter* sp., *Enterobacter* sp., *Pseudomonas* sp. and *Kilabsella* sp.) and Gram positive (*S. aureus, S. pneumonia*, Group A *Streptococcus* and *Bacillus anthrax*) all of which are causes of morbidity worldwide. There is extensive literature on the antibacterial effects of fresh garlic juice, aqueous and alcoholic extracts, lyophilized powders, steam distilled oil and other commercial preparations of garlic. The antibacterial effects of garlic have been studied by Reuter et al (**1996**) where they found that the active inhibitory principle of garlic is allicin or diallyl thiosulphinic acid.

2.1.1.4 CINNAMON Zeylanicum blume belongs to the botanical family of Lauraceae and was introduced in West Africa through the National herbarium of Ghana (University of Kumasi) in 1970. Cinnamomum zeylanicum is one of the oldest herbal medicines known, having been mentioned in Chinese texts as long as 4,000 years ago. It is often used for medicinal purposes due toits unique properties. Until now, morethan 300 volatiles were found as constituents ofessential oils of cinnamon. Jayaprakasha etal (**2002**) stated that the essential oil derived from cinnamon leaves is rich in eugenol, that from the roots in camphor and that from the buds shows a high amount of sesquiterpenes (α-bergamotene and α-copaene). It has been established that the oils and extracts from cinnamon possess a distinct antioxidant activity, which is especially attributed to the presence of phenolic and polyphenolic substances (**Jayaprakasha et al2006,Tomaino et al 2005**). These main propertiesof cinnamon are astringent, warming, stimulating, carminative, antiseptic, antifungal, antiviral, blood purifying, and aiding digestion. Eugenol and cinnamaldehyde are two very important terpenoids foundin cinnamon. Cinnamaldehyde and cinnamon oil vapors act as potent antifungal agents. The diterpenes found in the cinnamon oil have shown antiallergenic activity. The numerous uses of cinnamon as a medicinal herb imply the widespreadappreciation of its healing effects by herbalists worldwide. Unfortunately, often there is no scientific research to backup the claims that cinnamon does in fact have healing powers. Along with the medicinal effects come the side effects and interactions that medicinal cinnamon causes. Essential oil (EO) of *C. zeylanicum* Blume has shown varioustherapeutic actions (**Lawless 2002**) and is used as additive in tomatoes and spices soups aspreservative. Baratta et al (1998) have reported antimicrobial and antioxidant properties of the EO and its potential use in foods preservation. Simic et al (2004) have revealed its antifungal activities. Many studies revealed the apoptosis-inducing activity of some compounds nearcinnamon like taxol (**Kaufmann 1989, Cheng et al 2006**). Yehouenou et al(**2012**) found that Essential oils of cinnamon species have lot of biological activities including antimicrobial, antioxidant and antifungal properties. Furthermore, cytotoxic and apoptotic activities of several constituents were identified throughout its biological properties. Yehouenou et al (**2012**) also indicated the possibility of exploiting cinnamon essential oil in the fight against food-borne pathogens and adulterated microorganisms responsible for biodeterioration of stored foodstuffs. Jayaprakasha etal (**2002**) showed that the essential oil derived from cinnamon leaves is rich in eugenol, that from the roots in camphor and that from the buds shows a high amount of sesquiterpenes (α-bergamotene and α-copaene). It has been established that the oils and extracts from cinnamon possess a distinct antioxidant activity, which is especially attributed to the presence of phenolic and polyphenolic substances (**Jayaprakasha et al2006, Tomaino et al 2005**). In combination with other botanical products, *C. zeylanicum*showed potential clinical utility in patients with allergic rhinitis (**Corren et al 2008**). It exhibited insulin-mimetic action in 3T3-L1 adipocytes whileinhibiting the secretion of adiponectin (**Roffey et al 2006**) and decreased blood glucose levels in glucosetolerance test (GTT) (**Verspohl et al 2006, 2005**). It is also reported to have a role in wound healing (**Kamath et al 2003**). Additionally, it has insecticidal activity (**Abdel 2006**). The essential oil from *Cinnamomum zeylanicum* bark is rich in trans-cinnamaldehyde with antimicrobial effects against animal and plant pathogens,food poisoning andspoilage bacteria and fungi (**Mastura et al 1999**). *Cinnamomum zeylanicum* essential oil was tested for an effect on intestinal cell viability (**Fabian et al 2006**). Antioxidant activities of volatile extractsisolated from cinnamon were evaluated by various in vitro assays (**Lee and Shibamoto 2002**). The contents of glutathione (GSH) andlipid-conjugated dienes were studied in rats fed a high-fat diet along with cinnamon and it was reported that cinnamon stimulates the activity of antioxidant enzymes (**Dhuley 1999**). In addition, the effect of cinnamate, a phenolic compound found in cinnamon bark and other plant materials, on lipid metabolism and antioxidant enzyme activities in rats fed a high cholesterol diet has been studiedand indicated that cinnamon suppresses lipidperoxidation via the enhancement of hepaticantioxidant enzyme activities (**Lee et al 2003**). In a study of the wound healing action of an ethanol extract of cinnamon, the significant increase in wound healing was attributed to the antioxidant activity(**Kamath et al 2003**).

Cinnamon constituents possess antioxidant action and may prove beneficial against free radical damage to cell membranes (**Dragland et al2003**). Cinnamon has a marked antioxidant potential and may be beneficial in alleviating the complications of many illnesses related to oxidative stress in humans (**Ranjbar et al2006**). *Cinnamonum cassia* has been positively tested recently in type2 diabetics for its antidiabetic effect (**Khan et al 2003**). No dose-dependency of the effect could be derived, although the benefit to the patients was obvious (**Khan et al 2003**). Cinnamonum including other spices are well known to display an insulin-enhancing activity in vitro (**Khan et al 1990, Broadhurst et al 2000,Berrio et al 1992**). Aqueous extracts of Cinnamonum appear to increase glucose uptake and glucoseoxidation in vitro as well as glycogen synthesis and phosphorylation of the insulin receptor (**Imparl- Radosevich et al1998, Jarvill-Taylor et al 2001**); all these parameters improve the diabetic situation. Eugen et al (**2005**) reported that the *Cinnamomum cassia* extract has a direct antidiabetic potency. Barks of Cinnamomum plants are used as spice and herbal medicine, and *Cinnamomum zeylanicum* bark contains dimeric, trimeric, and higher oligomeric proanthocyanidins with doubly linked bis-flavan-3-ol units in the molecule. This class of proanthocyanidins is known to occur widely in common foods (**Gu et al 2004**) as well as the singly linked proanthocyanidins. Jayaprakasha et al (**2006**)isolated and identified non-volatile constituents and as well as antioxidant activities from cinnamon fruits. *Cinnamomum zeylanicum* essential oil showed stimulatory effects on macrophages. Phagocytosis and killing of invading microorganisms by macrophages constitute the body's primary line of defence against infections. *Cinnamomum zeylanicum* essential oil exhibits significant antioxidant activity in broiler chickens where it appears that this spice exerts antioxidant protection through its ability to activate the antioxidant enzymes and findings of that study established that*Cinnamomum zeylanicum*essential oil has appreciable immunostimulatory activity.*Cinnamomum cassia* is superior to *C. zeylanicum* and has an antidiabetic potency which should be used in the future (**Van Furt 1982**). Several investigations have been conducted related to antidiabetic activity of Cinnamomum zeylanicum bark and have shown positive effect. Cinnamon has been reported to have remarkable pharmacological effects in the treatment of hyperglycemia (**Kar et al 2003,Verspohl et al 2005**). Mukul et al (**2007**) reported that oral administration of ethanolic extract of*Cinnamomum zeylanicum* leaves in the doses of100, 150 and 200mg/kg body weight to rats significantly reduced their blood sugar level in allxon induceddiabetic rats under acute and sub acute studies. In their studys scientists demonstrated that intake of 1, 3, or 6g of cinnamon per day reduces serum glucose, triglyceride, LDL cholesterol and total cholesterol in people with type 2 diabetes which suggests that the inclusion of cinnamon in the diet of people with type2 diabetes will reduce risk factors associated with diabetes and cardiovasculardiseases. Tailang et al (**2008**) reported that *Cinnamomum zeylanicum*have antidiabetic, antioxidant and antimicrobial properties where they investigated its antiinflammatory activity using ethanol extract obtained from bark. Invitro and in vivo experiments were performed targeting TNF- using flow cytometry. Ethanol extract of *C. zeylanicum* showed suppression of intracellular release of TNF- in murine neutrophils as well as leukocytes in pleural fluid. The extract was found to inhibit TNF-gene expression in LPS-stimulated human PBMCs at 20µg/ml concentration. A potent anti-inflammatory activity of cinnamon extract is suggestive of its antiarthritic activity, which could be confirmed in various models of arthritis. In the antiinflammatory activity assay, it was demonstrated that the essential oil has a higher capacity to inhibit proIL-1beta protein expression induced by LPS-treated J774A.1 murine macrophage (**Tailang et al 2008, Chakrabarti et al 2006**).

2.1.1.5 ADHATODA VASICA Adhatoda is said to be non-poisonousto mammals, but kill fish, insects, and lower organisms.*A. vasica* is traditionally used in many of the following ways:

(1) juice from the leaves and the decoction of the leaves and roots are helpful in asthma, bronchitis and chronic coughs and breathlessness;

(2) used for bleeding due to idiopathic thrombocytopenic purpura, local bleeding due to peptic ulcer, piles, menorrhagia;

(3) relief in pyorrhoea and for bleeding gums by local application;

(4) relieves or eases muscular spasms, cramps or convulsions;

(5) stimulates contraction of the uterine muscle, facilitating or speeding up childbirth ;

(6) lowers blood pressure.

The leaves, the roots and flowers of*Adhatoda vasica* are extensively used in indigenous medicine as remedy for cold, cough, bronchitis and asthma. Both thedecoctions and powder from constituents of many preparations used in the ayurvedic medicine for various affections of the respiratory tract. In chronic bronchitis and asthma it is said to be very useful.*A. vasica*appears to possess some anthelmintic activity that may support the usage of this plant by local farmers to cure their animals in ethnoveterinary medicine. The leaves contain an alkaloid vasicine besides an essential. In ayurvedic medicine, *A. vasica*has been used for a multitude of disorders including; bronchitis, leprosy, blood disorders, heart troubles, thirst, asthma, fever, vomiting, loss of memory, leucoderma, jaundice, tumors, mouth troubles, sore-eye, fever, andgonorrhea. A decoction of the leaves of vasaka may be used to help with cough and other symptoms of colds. The soothing action helps irritation in the throat and the expectorant will help loosen phlegm deposits in the airway. It is useful in ridding the patient of coughing and asthma and can be given as a cure in any disease with which these symptoms are associated. Vasaka's special virtue is stopping bleeding due to the aggravation of pitta, through the mouth, nose, genitals, or the urinary systems. The leaves are dampened and then pounded, and one teaspoon of the resultant juice isuseful in cases of chronic bronchitis, asthma and tuberculosis. Boiled and puton the bladder region, the leaves produce a diuretic action, reduce the swelling of the kidney and lead to clearurination. Persons suffering from bleeding piles or diarrhea, accompanied by bleeding, or women suffering from menorragia can take the juice of vasaka 2 to 3 times a day to great advantage. Dried and powdered leaves also form a remedy for bronchitis in the dose of 40 grains twice a day. In liquid form the dose is about half to one teaspoon. In ancient times the root of the vasaka was tied on the back below the navel region and it is said that it produced safe, painless delivery. The leaves and roots containother alkaloids, vasicinone, vasicinolone and vasicol, which may contribute to the bronchodilatory effect through anticholinergic action on the vagal innervation of the bronchii. The bronchodilation effect is considerably increased after atropine administration. Studies have also shown vasa to be effective in the treatment of amlapitta (dyspepsia) and pyorrhea. Thein vitro growth of several strains of *Mycoplasma tuberculosis*was inhibited by the essential oil at concentrations in the range of 2-20ug/ml. Maurya and Jayendra (**2010**)stated that *Adhatoda vasica*has medicinal uses, mainly antispasmodic, fever reducer, anti-inflammatory, anti bleeding, branchodilator,antidibetic, antihelminthic, disinfectant, anti-jaundice, antiseptic, oxitocic and expectorant and has many other medicinal applications (**Shobha et al 2014**). In vitro sensitivity test with *A. vasica*leaf extract of microorganism of inflamed gingival showed significant antimicrobial activity. Josephin and Selva (**2012**) assessed the minimum inhibitory concentration (MIC) of *A. vasica* against clinical pathogen by disc diffusion method. The plant extract of*A. vasica* showed higher activity for different clinical pathogens in the order of

Klebsiella pneumonia >Staphylococcus aureus>Proteus valgaris>Pseudomonas aeruginosa > Streptococcus pyogens. Bhavani and Ballow (**2000**) stated that *A. vasica* has the antibacterial activity against Gram-positive and Gram-negative bacteria. The methanol extract of *A. vasica* exhibited high activity against the tested organisms rather than aqueous extract and showed the antimicrobial activity against *Staphylococcus aureus, Streptococcus pyogens, E. coli, Pseudomonasaeruginosa, Proteus vulgaris*and*Klebsiella pneumoniae* exhibiting the highest activity in methanol extract than the aqueous extract.The crude extracts obtained from the leaf of *A. vasica* using solvents of varies polarity such as ethanol, petroleum ether and water extracts exhibited the activity against*Staphylococcus aureus,Streptococcus pyogens, E. coli, Pseudomonas aeruginosa, Proteus vulgaris* and *Klebsiella pneumoniae*. From Josephin and Selva (**2012**) experimental result it was found that *A. vasica*showed the better activity in ethanol, diethyl ether and water against Gram-positive and Gram-negativeorganisms. The oil obtained from leaves flowers and roots of vasica plant possesses significantly high activity against tubercle bacilli.

2.1.2 MATERIALS AND METHODS

2.1.2.1 Study population (test organisms) Both Gram-positive (G+ve) and Gram-negative (G-ve) strains of bacteria were used as the test organism to observe the anti-bacterial activity of the compounds, which were*Staphylococcus aureus*(G+ve); and *Salmonella* sp., *Shigella* sp., *E. coli, Klebsiella* sp. (all are G-ve).

2.1.2.2 Selection, Collection and Size of Samples or plants The selected sample of different parts of plant, herbal, and spice materials were collected from different local areas and market in Dhaka city, Bangladesh. Five natural product materials were used in present study (**Table 2.1.1**). All sample containers were marked immediately and were transport to the laboratory with special polybag within 1h.

Table 2.1.1: Name of the different herbal plant and their parts used. [Ummey et al 2012, Zakaria et al 2012]

Name of the plants	Parts used
Trifala (Amla+ Haritaki+ Bibitaki)	Fruits
Java plum (*S. cumini*)	Fruits
Cinnamon (*C. verum*)	Bark
Henna (*L. inermis*)	Leaves
Cumin (*C. cyminum*)	Fruits

2.1.2.3 Test microorganisms*Bacillus subtilis, Bacillus cereus, Staphylococcus aureus, Bacillus subtilis, Bacillus cereus, Staphylococcus aureus* these organisms were collected from the Microbiology research laboraroty,Department of Microbiology, Primeasia University of Dhaka.The pure of which was previously collected from the Microbiology Department of Dhaka University.

2.1.2.4 Pure Culturepreparation With the help of a inoculating lop, the test organisms were transferred from the pure culture to the agar slants in a laminar unit, the incubated satins were then incubated at 37ºC for 18-24h to ensure the growth of test organisms. This culture was used for sensitivity test.

2.1.2.5Drying, Pulverization and Sample Preparation The fresh plant products were first washed with water to remove adhering dirt and then cut into small pieces, dried overnight in Hot air oven at 50°C. After complete drying, the entire portions were pulverized into a coarse powder with the help of a grinder machine and were stored in an airtight container for further use. 10g of the solid sample was weighed aseptically into a sterile jar and 90ml of distilled water was added to it. It was homogenized with sterile blender at 3000rpm for 5-10min. 1ml of homogenate was transferred to a test tube containing 9ml of sterile distilled water to make 10^{-1} dilution and shaken with vortex mixer. A serial dilution up to 10^{-8} was also made in the same procedure.

2.1.2.6Preparation of medicinal plant extracts and Standardization of inoculums In ethanolic process, the powdered 10g powders were added with 100ml distilled ethanol in a conical flasks. Then the solution was kept in room temperature for 24-72h, after that the homogenate was filtered with a sterile 11cm filter paper, then the extraction was kept in a sterilecontainer and stored at -20°C. In mechanical process, aqueous Trifala, Java plum, Cinnamon, Henna and Cumin extract was prepared according to methods previously reported by (**Hoque et al 2008**). 1g of residual ethanol extract powder was mixed with 10ml distilled water and these were kept in room temperature 2-3days. Then homogenate was filtered off with sterile cheese cloth and then 11cm filter paper. Filtrate was directly used for the sensitivity test. A standard stock of the bacteria isolates was prepared by suspending a loop full of each microbial growth in about 5ml of sterile saline. After incubation at 37°C for 12h, the turbidity was adjusted to be visually comparable with a 0.5 McFarland's standard giving a bacterial load of about 5×10^5cfu/ml (**Murray et al 1999**).

2.1.2.7 Preparation of extracts:

2.1.2.7.1 Mechanical process Aqueous extract of *A. sativum* and *P. emblica* were prepared according to methods previously reported by Onyeagba et al (**2004**). 100g of fresh, washed *A. sativum* cloves was macerated in a sterile, ceramic mortar.The homogenate was then filtered off with a sterile, muslin cloth and used directly for the sensitivity test. To ensure aseptic conditions, sterile gloves and face masks were worn and the entire extraction was carried out in a media room of the Microbiology Laboratory, Primeasia University, Dhaka.

2.1.2.7.2 Ethanolic process The powdered 10gm powders were added with 100ml distiled ethanol in a conical flask. then the solution was kept in room temperature for 24-72h, after that the homogenate was vaccume filtered with a sterial filter paper. Then the extraction was kept in a sterial container and stored at-20°C.

2.1.2.8 Standardization of inoculums (Murray et al. 2007) A standard stock of the bacteria isolates were prepared by suspending a loop full of each microbial growth in about 10ml of sterial saline. After incubation at 37°C for 12h, the turbidity was adjusted to be visually comparable with a 0.5 McFarland's standard giving a bacterial load of about 1.4×10^7cfu m/L.

2.1.2.9 Susceptibility test Suspensions of the bacterial isolates were made in sterile normal saline and adjusted to the 0.5 McFarland's standard. Each Nutrient agar (NA) agar plate was uniformly seeded by means of sterile spreader dipped in the suspension and spreaded on the agar plate surface, and the plates left on the bench for excess fluid to be absorbed (**Murray et al 2007**). Wells of 5mm in diameter, 4mm deep and about 2cm apart were punched in the NA agar with a sterile cork-borer. Approximately 100µl of the extracts were dropped into each well which filled them respectively to fullness. The setup was allowed to stabilize for 3h before being incubated at 37°C for 24h as described previously by Shahidi (**2004**), Murray et al (**2007**) and Aibinu et al (**2007**).

2.1.2.10Preparation of Test plates and Agar Well Preparation Suspensions of the bacterial isolates were made in sterile normal saline and adjusted to the 0.5 McFaeland's standard. Each nutrient agar plate was uniformly seeded by means of sterile cotton swab dipped in the suspension and spread on the agar plate surface, and the plates left on the bench for excess fluid to be absorbed (**Murray et al 1999**). Agar Wells of 6mm in diameter, 4mm deep and about 2cm apart were punched in the NA with a sterile cork-borer. Approximately 80µl of the extracts were dropped into each well which filled them respectively to fullness. The setup was allowed to stabilize for 3h before being incubated at 37°C for 24h as described previously (**Murray et al 1999, Aibinu et al 2007**). In present study, the aqueous extract concentration of 100µl/well was used.

2.1.2.11Determination of MIC- MBC and zone of inhibition (ZOI) in different concentration

The MIC is the lowest concentration of the test sample or drug at which it shows the highest activity against microorganism. The minimum inhibitory concentration was determined by serial dilution techniques using nutrient broth media. The minimum inhibitory concentration of pure plant was determined against five all of the selected microbes. Dilutions showing no visible growth for the MIC was sub-cultured and incubated at 37°C for 24h. The lowest concentration of the extracts yielding no growth on the NA plate wasrecorded at the MBC. Serial dilutions 1:2, 1:4, 1:8, 1:16, 1:32 and 1:64 mg/ml ($^{v}/_{v}$ in case of aqueous extract) and ($^{w}/_{v}$ in case of ethanolic extract) were prepared from stock solutions. Then their antimicrobial activities were determined by agar well diffusion procedure.

2.1.3 RESULT

2.1.3.1 Antimicrobial activity It has been observed from that the aqueous extract of Triphala, Java plum, Henna, Cumin and Cinnamon had good antimicrobial activity against all the selective microbes except *Klebsiella* sp. But the aqueous extract of Cinnamon had not any antimicrobial activity against *S. aureus*. Whereas Cumin showed less sensitivity and Triphala, Java plum, Henna and Cinnamon showed higher antimicrobial activity. It was found that *Klebsiella* sp. was resistance to all the ethanolicextracts except in cinnamon (**Table 2.1.2**). Different concentrations of all the samples were applied in different selected microbes (1.4×10^{7}cfu) and the ZOI assayed (**Table 2.1.3**). It was observed that mor or less 1:8 concentration showed better result and in the 1:16, the ZOI started to decreased and 1:64 it showed almost negative result. In case of ethanolic extraction, Triphala, Java plum, Henna, Cumin and Cinnamon were tested for their antimicrobial activity against all of the selected microbes. The ethanolic extract was neutralized for the determination of the effect of pH (ethanol) on antimicrobial activity.

Table 2.1.2: Sensitivity and resistance pattern of the organism against test products. [Ummey et al 2012, Zakaria et al 2012]

Name of organisms	Name of products									
	Triphala		Java plum		Henna		Cumin		Cinnamon	
	S	R	S	R	S	R	S	R	S	R
Salmonella sp.	+	_	+	_	+	_	+	_	+	_
Shigella sp.	+	_	+	_	_	+	+	_	+	_
S. aureus	+	_	+	_	+	_	+	_	_	+
E. coli	+	_	+	_	+	_	_	+	+	_
Klebsiella sp.	+	_	+	_	+	_	+	_	_	+

*S= Sensitive; R= Resistant

Table 2.1.3: Determination of antimicrobial activity of Triphala, Java plum, Henna, Cumin and Cinnamon in different concentration against different microbes. [Ummey et al 2012, Zakaria et al 2012]

Different Concentration	Name of the Product	Name of the Organisms (ZOI)				
		Salmonella sp.	*Shigella sp.*	*S. aureus*	*E. coli*	*Klebsiella sp.*
	Triphala	35	23	29	25	30
Neat	Java plum	30	21	25	20	29
	Henna	21	-	19	18	19
	Cumin	13	-	14	12	11
	Cinnamon	11	13	-	12	-
	Triphala	26	19	21	19	22
1:2	Java plum	27	17	19	16	23
	Henna	16	-	13	12	14
	Cumin	9	-	10	9	8
	Cinnamon	7	9	-	8	-
	Triphala	21	14	16	13	18
1:4	Java plum	21	11	12	12	17
	Henna	10	-	8	8	10
	Cumin	4	-	7	5	5
	Cinnamon	5	4	-	5	-
	Triphala	16	9	11	9	14
1:8	Java plum	17	7	8	8	13
	Henna	5	-	4	5	4
	Cumin	0	-	4	0	0
	Cinnamon	0	0	-	0	-
	Triphala	12	4	7	4	10
1:16	Java plum	13	4	4	3	9
	Henna	0	-	0	0	0
	Cumin	0	-	0	0	0
	Cinnamon	0	0	-	0	-
1:32	Triphala	8	0	4	0	7
	Java plum	9	0	0	0	4
1:64	Triphala	0	0	0	0	0
	Java plum	0	0	0	0	0

Aqueous and ethanolic extract of *A. sativum* and*P. emblica,* were tested (well method) for their antimicrobial activety against all of the selected microorganisms. In this antibacterial screening of crude extract were used at a concentration of 100μg/well. **Table 2.1.4** shows zone of inhibition of *A. sativum* and *P. emblica* (crude extract). It was observed that the aqueous extracts of *A. sativum* showed remarkable activity against all the tested organisms with the highest activity found on *Psudomonas* sp.(38mm zone diameter) at 100μg/well concentration. Zone of inhibition showed against *A. sativum* were satisfactory antimicrobial activity. Onyeagba et al (**2006**) worked on the same (*A. sativum*) and found no zone of inhibition against *A. sativum*. This is may be due to the species difference or the *A. sativum* difference in different biologic condition. Aqueous extract of *P. emblica* was also showed a good activity against all the tested organisms. De Britto et al (**2011**) showed satisfactory activity of *P. emblica* against *Xanthomonas* sp., *Aeromonas* sp. and *Campestris hydrophila* sp., but in present study it was more effective result was found aginst all the selected microbes. On the other hand, the ethanol extract of A. sativum (**Table 2.1.5**)is also active against the test organisms with highest activity on *S. aureus*(17mm) at 100μg/disc, but Yusha'u et al (**2008**) observed that ethanol extract of*A. sativum* showed highest (51mm) activity on *E. coli*at 200μg/disc (Habib et al (**2007**) reported sensitivity pattern of different microbes against chloroform extraction of*P. emblica* where no sensitivity was found whereas inpresent study, zone of inhibition was observed against all the tested organimsms. This difference of observation between Habib et al (**2007**) and present study may be due to species differences or ethanolic effect.

Table 2.1.4: Determination of antimicrobial activity of *A. sativum* and *P. emblica* (Crude Extract). [Ummey et al 2012, Zakaria et al 2012]

Name of organisms	Zone of inhibition (mm)	
	A. sativum	*P. emblica*
Psudomonas sp.	38	20
S. typhi	28	22
S. aureus	24	24
E. coli	27	18
Klebsiella sp.	36	19
S. paratyphi	30	21
B. subtilis	28	20
Salmonella sp.	39	19
B. cereus	22	22

Table 2.1.5: Determination of antimicrobial activity of *A. sativum* and *P. emblica* (Ethanolic Extract). [Ummey et al 2012, Zakaria et al 2012]

Name of organism	Zone of inhibition (mm)	
	A. sativum	*P. emblica*
Klebsiella sp.	11	15
B. subtilis	12	16
B. cereus	6	11
Psudomonas sp.	17	15
S .typhi	10	9
E. coli	0	0
S. paratyphi	0	9
S. aureus	17	10

2.1.3.2 MIC and MBC It was observed that the MIC of Triphala against *Salmonella* sp., and *E. coli* were 12.5µg/ml; whereas MIC against *Shigella* sp. and *S. aureus* were 50µg/ml and 25µg/ml, respectively. The MBC of Triphala against *Salmonella* sp., and *E. coli* were 50µg/ml and 100µg/ml against *Shigella* sp., *S. aureus* were observed all other organisms. It was observed that the MIC of Triphala against *Salmonella* sp. was 6.25µg/ml whereas MIC against *Shigella* sp., *E. coli, Klebsiella* sp. were 12.5µg/ml; *S. aureus*25.0µg/ml. The MBC of Triphala against *Shigella* sp., *S. aureus, E. coli* and *Klebsiella*sp. were 100µg/ml and 50µg/ml MBC was observed against all other organisms. In case ofJava plum, MIC against *Shigella* sp., *S. aureus* and *E. coli*, were 25 µg/ml, MIC 6.25 µg/ml to *Salmonella* sp.; and MIC of 12.5µg/ml were observed in case of other microbes (**Fig. 2.1.1**). On the other hand, MBC of 100µg/ml were observed against all other microbes except *Salmonella* sp. where it was 50µg/ml MBC. The MIC of Henna against *S. aureus, E. coli*, were 50µg/ml whereas *Salmonella* sp.,*Klebsiella* sp. were 12.5 and 25µg/ml respectively. MBC was found 100µg/ml against *S. aureus, E. coli* and *Klebsiella* sp. were found 100µg/ml and 50µg/ml MBC was against *Salmonella* sp. In case of Cumin, 50µg/ml of MIC was against *Salmonella* sp., *S. aureus* and *Klebsiella* sp. observed whereas against*Shigella* sp. MIC was 25µg/ml. MBC was found 100µg/ml against all microbes. 50µg/ml of MIC was against *Salmonella* sp., *Shigella*sp., and *E. coli* observed whereas the MBC was found 100µg/ml against all selected microbes against Cinnamon (**Fig. 2.1.2**).

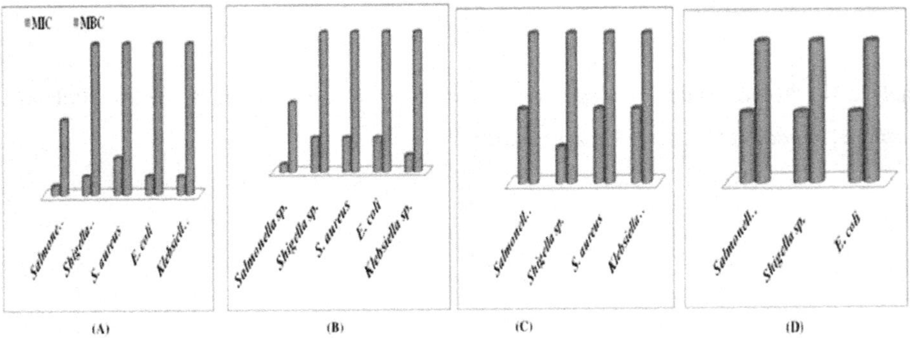

Figure 2.1.1: Determination of MIC and MBC (natural extract) of (A) Triphala (B) Java plum (C) Cumin (D) Cinnamon against different microbes. [Ummey et al 2012, Zakaria et al 2012]

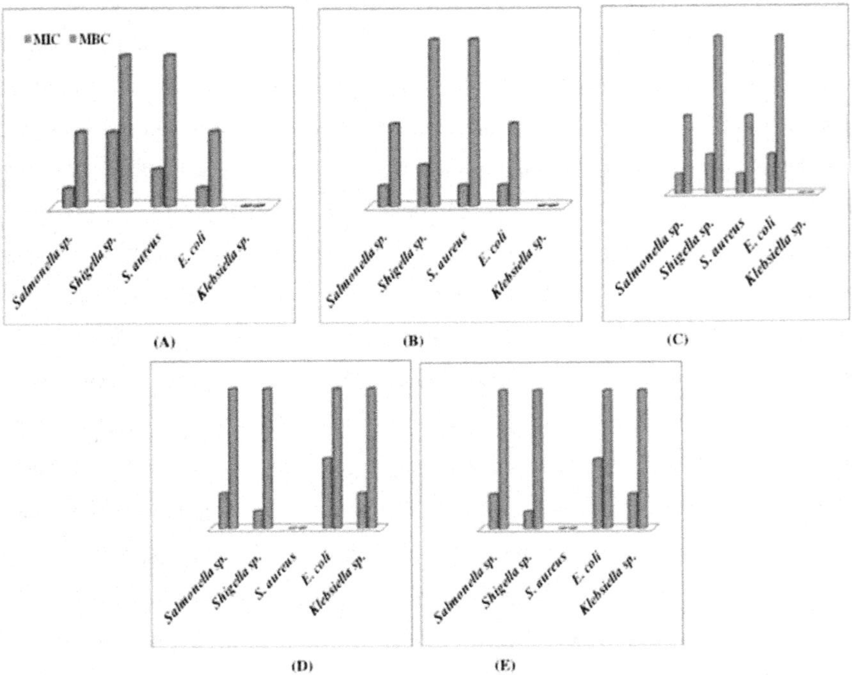

Figure 2.1.2: Determination of MIC and MBC (ethanolic extract) of (a) Triphala (b) Java plum (c) Henna. (d) Cumin and (e) Cinnamon against different microbes. [Ummey et al 2012, Zakaria et al 2012]

It was observed that the MIC of *A. sativum* against *Salmonella* sp., *S. typhi, B. cereus* and *S. aureus*was12.6µg/ml whereas MIC against *Klebsiella* sp., was 3.12*S. paratyphi*, MIC against *B. subtilis* and *Psudomonas*sp.and *E. coli* were 6.25µg/ml.The MBC of *A. sativum* against *Salmonella* sp. and *S. paratyphi* were 100µg/ml, 25 µg/ml MBC was observed against *Klebsiella* sp. and *E.coli*and 50 µg/ml MBC was observed against all other organisms (**Fig 2.1.3**). Whereas, the MIC of *A. sativum* (ethanolic extract) against *Klebsiella* sp. *Salmonella* sp. and *S. aureus* were 50.0µg/ml and *Psudomonas* sp. *S. typhi, B. cereus, B. subtilis* were 25.0µg/ml. MBC were found 100µg/ml against all microbes (**Fig 2.1.4**). In **Figure 2.1.5**, it was observed that the MIC of *P. emblica* against *Salmonella* sp., *S. typhi, B. cereus* were 12.6 µg/ml whereas MIC against *Klebsiella* sp., *S. paratyphi*,MIC against *B. subtilis* and*Psudomonas, S. aureus* and *E. coli*were 6.25µg/ml.The MBC of *P. emblica* against *Salmonella* sp. and *S. paratyphi* were 100µg/ml, 25 µg/ml MBC was observed against *E. coli*and 50 µg/ml MBC was observed against all other organisms. It was observed that the MIC of *P. emblica* (ethanolic extract) against*Salmonella* sp., *B. cereus, B. subtilis* and*S. aureus*were 12.5µg/ml; whereas MIC against *S. typhi, Psudomonas* sp. and *Klebsiella* sp. were 25.0µg/ml. The MBC of A. *sativum* against *Salmonella* sp., *Klebsiella* sp. *Psudomonas* sp. and *S. paratyphi* were 100µg/ml and 50µg/ml MBC was observed against all other organisms (**Fig 2.1.6**). Fani et al (**2007**) studied on MIC of*A. sativum*against *S. mutans* ranged from 4-32mg/mL which was slightly lower compared with our study where *A. sativum*(aqueous extract) showed MIC from 6.25 to 25µg/ml. This may be due to the species differences. Owhe-Ureghe et al (**2010**) observed that the MIC of *A. sativum* against the test isolates ranged from 31.25-62.5 mg/mL whereas in present study, MIC of *A. sativum*ranges from 6.25-25.0 µg/ml in case of aqueous exrtracts.

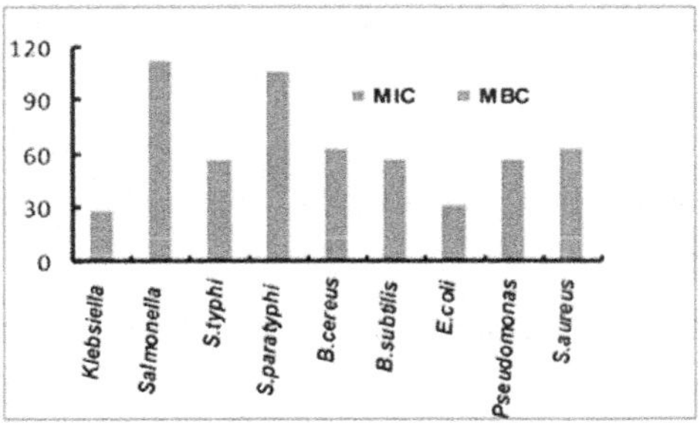

Figure2.1.3- Determination of MIC and MBC of A. sativum (crude extract). [Ummey et al 2012, Zakaria et al 2012]

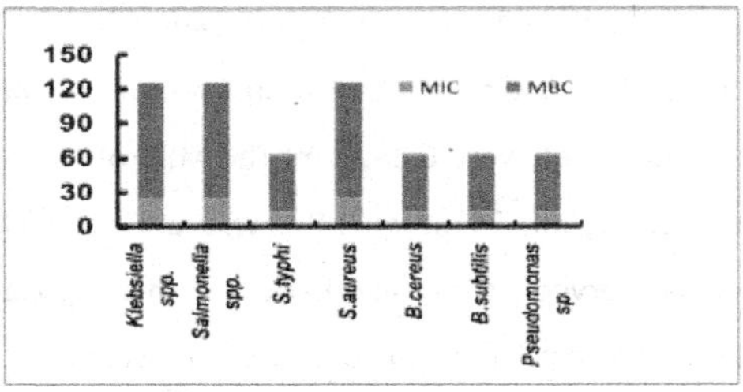

Figure2.1.4- Determination of MIC and MBC of A. sativum (Ethanolic extract). [Ummey et al 2012, Zakaria et al 2012]

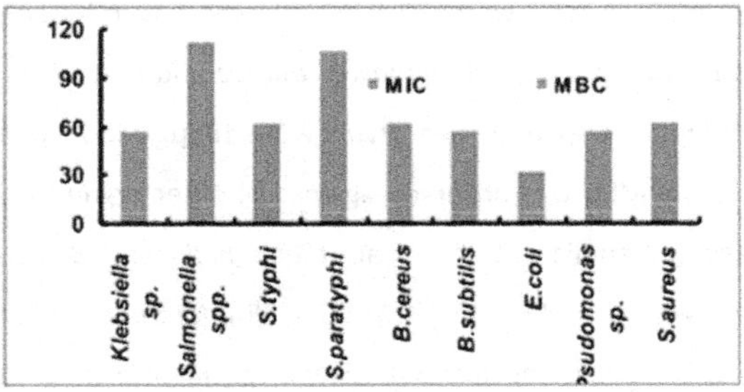

Figure 2.1.5 - Determination of MIC and MBC of P. emblica (crude extract). [Ummey et al 2012, Zakaria et al 2012]

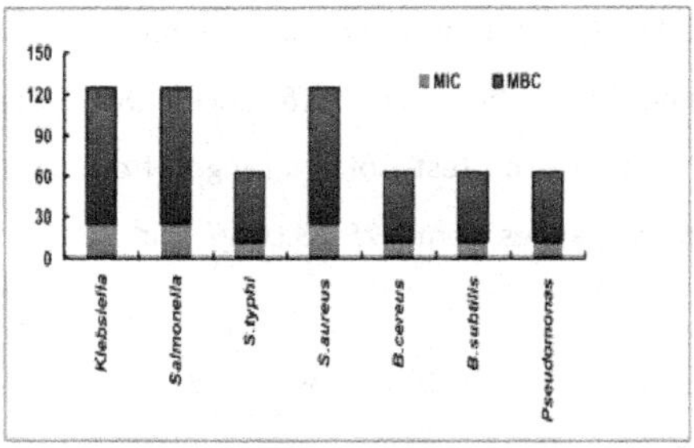

Figure 2.1.6 - Determination of MIC and MBC of P. emblica (Ethanolic extract). [Ummey et al 2012, Zakaria et al 2012]

2.1.3.3 Comparison with commercial antibiotics In case of *Klbesiella* sp., highest zone of inhibition was observed 38mm against *A. sativum* (aqueous). Clear zone of inhibition of 19mm was found in case of*P. emblica*(aqueous); whereas 29, 25 and 17mm clear zone were observed against Tetracycline, Ciproflaxacin and Cholamphenicol, respectively. In case of *Psudomonas* sp. 39 and

21mm clear zones were showed against *A. sativum* (aqueous) and *P. emblica* (aqueous) respectively whereas clear zone by choramphenicol and ciprofloxacin were 23mm. At the time of *S. typhi* clear zones were 30 and 22mm against *A. sativum* (aqueous) and *P. emblica* (aqueous), respectively. 28 and 20 mm clear zones were showed against Cholamphenicol and Ciproflaxacin (**Fig2.1.7**). In case of *E.coli* clear zone of inhibition of 20 and 29mm was observed against *P. emblica* (aqueous) and *A. sativum* (aqueous), respectively; whereas zone of inhibition by Ciproflaxacin, Cefotaxime, Choramphenicol and Gentamicin were 30, 29, 26 and 28, respectively against *E.coli*. In case of*B. subtilis*, zone of inhibition 32mm was found against *A. sativum*, at the same time, zone of inhibition of 30mm was observed against Ciproflaxacin. In case of B. cereus *A. sativum* (aqueous) and *P. emblica* (aqueous) showed34 and 20mm zone of inhibition. Clear zone by Ciproflaxacin, Merupenam and Vancomycin were 30, 30 and 20mm, respectively (**Fig2.1.8**). Satisfactory results were observed only in case *A. sativum* and*P. Emblica*(aqueous extract) as compared with commercial antibiotics. Sometimes*A. sativum* was better than the commercial one. Extracts of *P. emblica, A. sativum* and showed that they are effective against all the tested pathogens *Psudomonas* sp., *S. aereus, Salmonella* sp., *S. typhi, S. paratyphi, B. cereus, B. subtilis, E. coli, S. aureus* in controlling their growth in vitro in culture condition. They all have a bacteristatic and bactericidal activity when tested in vitro using crude preparation. All of the plant products (Aqueous and ethanolic) *A. sativum* and *P. emblica* (aqueous extract) were best they both could be able to inhibit the growth of all the sensitive microbes. Here the microbes we selected for our experiment can cause different types of diseases.As aqueous extracts of *A. sativum* and *P. emblica* are more sensitive to the microbes.Although in some cases *A. sativum* was more effective than commercial antibiotics like tetracycline and chloramphenicol. gallic and *P. emblica* may be used successfully for treating various diseases causing microbes agent *S. aureus., Salmonella, E. coli, S. aureus* etc. Further *in vivo* studies are necessary.

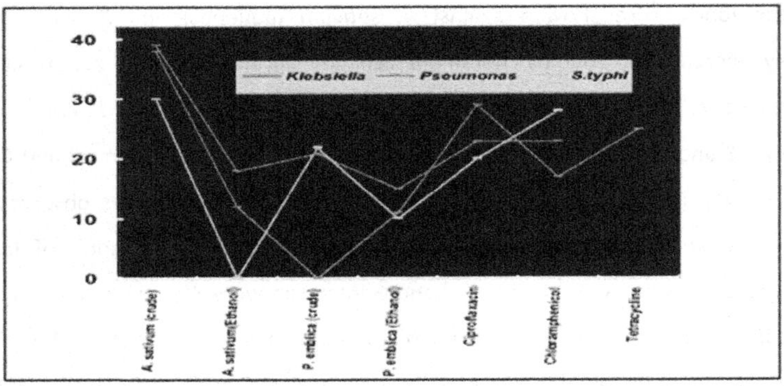

Figure2.1.7: Comparative analysis of *A. sativum* and *P. emblica* withcommercial antibiotics. [Ummey et al 2012, Zakaria et al 2012]

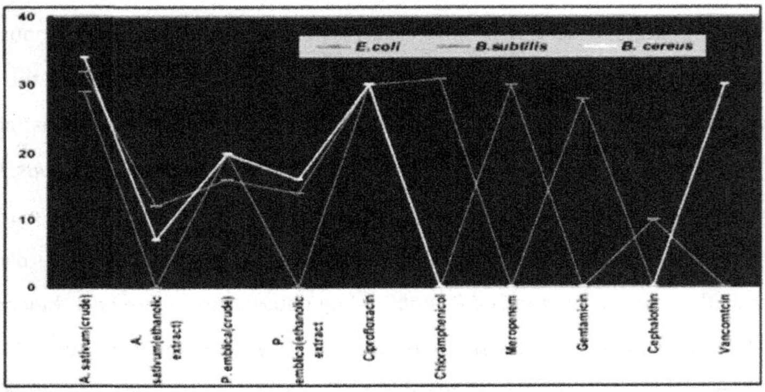

Figure2.1.8: Comparative analysis of A. sativum and P. emblica withcommercial antibiotics

2.1.3.4Combined effect In natural, Java plum+Cumin showed satisfactory results against all the selected microbes except *Klebsiella* sp. (**Fig.2.1.9**). In case of ethanolic extracts, Java plum+Cumin and Triphala+Cinnamon showed satisfactory results against all the selected microbes (**Fig.2.1.10**). Only, *Klebsiella* sp. which did not show better result in case of Java plum+Cumin. A Comparative analysis of ZOI of neutral and crude ethanol extraction against test organisms was also done to know the efficacy ofthese two extracts (**Table 2.1.6**).

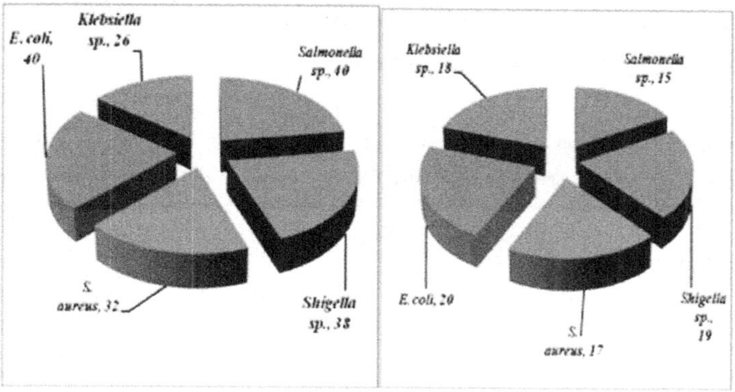

(a) (b)

Figure 2.1.9 Combined antimicrobial effect (natural extract) of (a) Java Plum+Cumin and (b) Triphala+Cinnamon against different microbes. [Ummey et al 2012, Zakaria et al 2012]

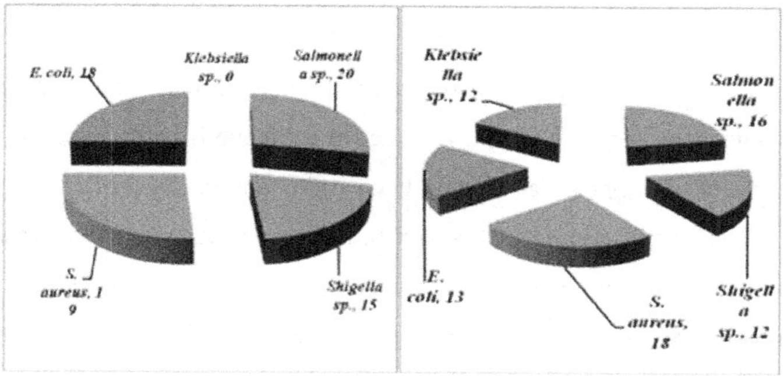

(a) (b)

Figure 2.1.10 Combined antimicrobial effect (ethanol extract) of (a) Java plum+Cumin and (b) Triphala+Cinnamon against different microbes. [Ummey et al 2012, Zakaria et al 2012]

Table 2.1.6: Comparative study of ZOI of neutral and crude ethanol extraction against test organisms. [Ummey et al 2012, Zakaria et al 2012]

	ZOI (mm)									
Name of	Triphala		Java plum		Henna		Cumin		Cinnamon	
organisms	Neutral	Crude	Neutral	Crude	Neutral	Crude	Neutral	Crude	Neutral	Crude
Salmonella sp.	15	26	20	27	11	21	14	24	11	20
Shigella sp.	10	17	11	17	10	18	9	18	13	20
S. aureus	12	22	12	22	9	13	11	22	0	0
E. coli	9	16	10	18	11	17	10	17	9	15
Klebsiella sp.	0	0	0	0	0	0	0	0	10	17

2.1.4 DISCUSSION

The screening and isolation of bioactive compounds from plants is increasing, as these compounds are much safer and less toxic than the chemically synthesized drugs. The increasing trends in emergence of resistance among infectious agents, treatment failures and recurrences of infection had led to the search of biologically active compounds or products from plants. The antibacterial activity of Triphala has already been documented. The extract of triphala shows antimicrobial activity against some of the oral bacterial species tested (**Srikumar et al 2007**). Hence the study was done to evaluate the antibacterial effects of Triphala, Java plum, Henna, Cumin and Cinnamon against some of the bacterial pathogens. The aqueous and ethanolic extract of seeds of Java plum was tested for their activity against five different species. The extract showed great activity against *Salmonella* sp., *Shigella* sp., *S. areus* and *E. coli* but *Klebsiella* sp. whereas was inhibited. The aqueous extract of the entire sample showed more inhibition than the ethanolic extracts against all the species of bacteria that were inhibited. The aqueous extract of samples showed more inhibition than the ethanolic extracts against all the species of bacteria that were inhibited. Antimicrobial effect of aqueous extract of samples were higher than the against *Salmonella* sp. The antimicrobial effect of Triphala may be due to the presence of secondary metabolites, one of the compounds, Gallic acid, Chebulagic acid and Chebulinic acid was found to possess antimicrobial effect on oral bacteria's. Hence further analysis of the compounds and their testing against microorganisms would help to develop a new antimicrobial agent that may be of great importance. The results for the antibacterial screening have shown that all the extracts in this present study have antibacterial activity. The results of the inhibition of bacterial growth have shown that the extracts are active at high concentration and inactive at very low concentrations. Thus the study may suggest that the inhibition of bacterial growth activity of the extracts is dose dependent. It was observed that the aqueous extracts of triphala showed remarkable activity against all the tested organisms with the highest activity found on *Salmonella* sp. (35mm zone diameter) at 100µg/well concentration. ZOI showed against triphala was satisfactory antimicrobial activity. Srikumar et al (**2007**) worked on the same (triphala) and found no ZOI against triphala. This ismay be due to the species difference or the triphala difference in different biologic condition. Aqueous extract of Java plum was also showed a good activity against all the tested organisms. Aparna et al (**2015**) showedsatisfactory activity of Java plum against *S. aureus, Staphylococcus saprophyticus, E. coli,Pseudomonas aeurioginosa* and *Proteus vulgaris*. But in present study it was more effective result was found against all the selected microbes. In case of henna, cumin, it showed antimicrobial activity against all the selected microbes but the ZOI were not satisfactory like Triphala and Java plum. On the other hand, the ethanol extract of Triphala is also active against the test organisms with highest activity on*Salmonella* sp. (26mm) at 100 µg/disc, but (**Srikumar et al 2007**) observed that ethanol extract of triphala showed highest (16mm) activity on *S. aureus, V. cholera*, and at *E. coli* at 0.1µg/ml. In case of Java plum satisfactory results were found against all the microbes which demonstrate almost same results as the observation by Cowan (**1999**) where pronouncedactivity of Java plum was demonstrated against ethanolic extract where The MBC of the *S. cumini*extract was 62.5µg/ml for the bacteria*S. aureus*. This may be due to the species difference or difference in different biologic condition. Ibukun et al (**2007**) reported sensitivity pattern of different microbes against ethanol extraction of triphala were *E. coli*33mm,*Klebsiella* sp. 30mm zone diameter, whereas in present study, no ZOI was observed against*Klebsiella* sp. but other tested organisms. This difference of observation between and present study may be due to heating of triphala in experiment or species differences. *Salmonella* sp. and *S. aureus* showed better effective results in of cumin against different microbes than the present study which might be due to species difference or concentration variation. In case of combination of Java plum+Cumin showed ZOI was 14mm against *Shigella* sp. and *Klebsiella* sp., whereas in present study, range of ZOI against the selected microbes was observed from 14-22mm (**Ibukun et al 2007**). Solvent extracts of Triphala and Java plum showed a remarkable antibacterial activity on almost all test organisms. In this study, triphala (aqueous extract) showed MIC from 6.25 to 25μg/ml. on the other hand,Gram-positive bacteria were more susceptible to the inhibitory effects of the plant extract. Best activity sensitivity showed against *Salmonella* sp. in aqueous and ethanol extract. *S. aureus* showed best sensitivity in ethanol extract and *Klebsiella* sp. showed no sensitivity in all extract. The antibacterial activity attribute to its chemical constituent as reported by many researches such as Hydroxychavicol. Phenolic compounds such as cavicol, cavibetol, carvacrol, eugenol, and allipyrocatecho (**Fabricant and Farnsworth 2001, Beuchat and Golden 1989**). In this experiment G+ve bacteria more sensitive than G-ve bacteria and show this result for cell membrane permeability. G+ve bacteria have single layer and lack the natural sieves effect against large molecule. On the other hand, G-ve bacteria have multi – layered and complex cell wall structure (**Hawkey 1998**). Organic extracts were more potent than aqueous extract for two reasons: because of the nature of biological active components enhanced in the presence of solvent such as stronger extraction capacity of solvent produced greater number of active constituents responsible for the antibacterial activity; and betel leaf extract is rich in less polarity phenol compound (ethanol and metabolic) which were more efficient in extracting the active compounds from this species due to its no polar components (**Brković et al 2006, Cowan 1999**). Low/negligible antimicrobial action of aqueous extracts ascribed to low level of the anionic components (**Sukanya et al 2009**). No inhibition was observed with negative controls, which proves that solvents could not act as antibacterial agents. MIC value for G-ve>G+ve bacteria; higher dose of Gram-posiative bacteria will be required to cause infection. *Staphylococcus aureous* were found to be most sensitive (MIC: 0.78mg/l); others>more resistant (>1.25mg/l). MIC<MBC value; plant extracts were bacteriostastic at lower concentration and bactericidal at higher concentration. Low MIC exhibited by this plant extract showed that they can be used as an alternative to orthodox antibiotics. No inhibition was observed with controls, which proves that solvents could not act as antibacterial agents. All the test organisms were multi drug resistant. There are all organisms are multidrug resistant means drug (antibiotic) not work against bacteria, but this two natural product work against bacteria. Combined effect of Triphala and Cinnamon was also good but not as good as combination of Java Plum and Cumin showed. The highest ZOI was 20mm given by *E. coli*and the lowest ZOI was 15mm given *Salmonella* sp. Extracts of riphala,

Java plum, Henna, Cumin and Cinnamon showed effective antimicrobial agent against all the tested pathogens in controlling their growth in vitro in culture condition. They all had a bacteristatic and bactericidal activity against broad spectrum of microorganisms (Gram-positive and Gram-negative bacteria) when tested in vitro using crude preparation. All of the natural plant products (aqueous and ethanolic); Triphala and Java plum (aqueous extract)were found best antimicrobial activity and both of them could be able to inhibit the growth of all the sensitive microbes. In case of ethanolic extracts of these natural plant products, it was interestingly found that they had antimicrobial activity in all the cases. The microbes that were selected in this present study can cause different types of diseases. Although in some cases, there is need fordetailed scientific study of traditional medical practices to ensure that valuable therapeutic knowledge of some natural plants is preserved and also to provide scientific evidence for their efficacies. Also another study will be needed to establish the exact component or pharmacological standardization and clinical evaluation in Triphala.

2.1.5 CONCLUSION

Recent trends and developments in the area of animal nutrition have been characterized by an increasing interest in the potential impact of plants, herbs and spiceson the immune function and antioxidant status of humans and animals. Research activities have been intensified; however, understanding of the specific mode of action and the functional aspects of phytogenic additives still needs many more in depth studies. These medicinal herbs constitute indispensable components of the traditional medicine practiced worldwide due to the low cost, easy access and ancestral experience. The antibacterial and phytochemical screening insured the importance of these plants and the intensive use by the healers as traditional medicines and pointed a good guiding for further research studies in these plants.As the global scenario is now changing towards the use of nontoxic plant products having traditional medicinal use, development of modern drugs from Java plum should be emphasized for the control of various diseases. In fact, time has come to make good use of centuries-old knowledge on Triphala and java plum through modern approachesof drug development. Locally available and can be easily cultivated a good choice for the development of new strategies for therapy of bacterial diseases. Quite a significant amount of research has already been carried out during the past few decades in exploring the chemistry of Triphala, java plum, henna, cumin and cinnamon.

2.1.6 REFERENCES

Anushia C, Sampathkumar P and Ramkumar L. 2009. Antibacterial and antioxidant activities in cassio auriculata. Global J. Pharm. 3(3): 127-130.

Akhtar MS and Iqbal J. 1991. Evaluation of the hypoglycaemic effect of Achyranthes aspera J Ethnopharmacol. 31: 49.

Ankri S and Mirelman A. 1999. Antimicrobial properties of allicin from garlic. Microb. Infect. 2: 125-129.

Anonymous 1985. The Wealth of India, Raw Materials. New Delhi: Publication and Information Directorate, Center for Scientific and Industrial Research (CSIR), p. 57.

Abdel HAS 2006. Efficacy of Cinnamomum zeylanicum on third stage larvae and adult fecundity of Musca domestica and Anopheles pharoensis. J. Egypt Soc. Parasitol. 38: 475-482.

Ayurvedic Formulary of India. 2003. Part I, 2nd English ed. New Delhi: Controller of Publications.

Bhattarai N. 1994. Folk herbal remedies for gynaecological complaints in Central Nepal. Int J Pharmacog. 32(1):13-26.

Brindha P, Sridharan G, Pradeep V and Sasikumar S. 2008. Antitumor activity and in vivo antioxidant status of Hyptis suaveolens against Ehrich ascites Carcinoma in Swiss albino mice. Indian Drug, 45:10.

Bhattacharya A, Chatterjee A, Ghosal S and Bhattacharya SK. 1999. Antioxidant activity of active tannoid principles of Emblica officinalis (amla). Indian J Exp. Biol. 37(7): 676-80

Bhavani SM and Ballow CH. 2000. New agents for gram-positive bacteria. Current opinion in Microbiol., 3: 528-34.

Baratta MT, Dorman HJD, Deans SG, Figueredo AC, Barroso JG and Ruberto G. 1998. Antimicrobial and antioxidant properties of some commercial essential oils, Flavour Frag J. 13(4): 235-244.

Beuchat LR and Golden DA. 1989. Antimicrobials Occurring Naturally in Foods. Food Technology, 43, 134-42.

Brkovi ć LD, Ljiljana Č, Slavica SS. 2006. Antibacterial activity of some plants from family Apliaceae in relation to selected phytopathogenic bacteria. Kragujevac J. Sci. 28, 65-72.

Berrio LF, Polansky MM, Anderson RA. 1992. Insulin activity: stimulatory effects of cinnamon and Brewer's Yeast as influenced by albumin. Horm Res 37: 25–29.

Broadhurst CL, Polansky MM, Anderson RA. 2000. Insulin-like biological activity of culinary and medicinal plant aqueous extracts in vitro. J Agric Food Chem 48: 849–852.

Cowan MM. 1999. Plant products as antimicrobial agents. Clin Microbiol Rev. 12(4): 564-82.

Chakrabarti S, Zee JM, Patel KD. 2006. Regulation of matrix metalloproteinase-9 (MMP-9) in TNF-stimulated neutrophils: novel pathways for tertiary granule release. J. Leukoc. Biol. 79: 1-9.

Corren J, Lemay M, Lin Y, Rozga L, Randolph RK. 2008. Clinical and biochemical effects of a combination botanical product (ClearGuard) for allergy: a pilot randomized double-blind placebo-controlled trial. Nutr. J. 7: 20.

Cheng SS, Liu JY, Hsui YRand Chang ST. 2006. Chemical polymorphism and antifungal activity of essential oils from leaves of diffent provenances of indigenous cinnamom (Cinnamomum osmophoeum), Biores. Technol. 97: 306- 312.

Chopra, RN, Nayar SL and Chopra IC. 1986. Glossary of Indian medicinal plants. Council of Scientific and Industrial Research, New Delhi. 185-25.

Chaudhuri RK. 2004.Standardized extract of *Phyllanthus emblica*. A skin lightener with anti-aging benefits. Proceedings PCIA Conference, Guangzhou, China 9-11.

Chungsamarnyart N, Jiyajinda S and Jangsawan W. 1991. Larvicidal effect of plant crude extracts on the tropical cattle tick (Boophilus microplus). Kasetsert J. 25:80-89.

Calixto JB, Antos ARS, Filho VC and Tunes RA. 1998. A reviewof the plants of the genus phyllanthus: their chemistry, pharmacology, and therapeutic potential. J. Med. Bio. 31:225-258.

Dhuley JN. 1999. Antioxidant effects of cinnamon (Cinnamomum verum) bark and greater cardamom (Amomum subulatum) seeds in rats fed high fat diet. Indian J Exp Biol 37: 238-242.

Dragland S, Senoo H, Wake K, Holte K, Blomhoff R.2003. Several culinary and medicinal herbs are important sources of dietary antioxidants. J Nutr 133: 1286-1290.

Deresse D. 2011. Antibacterial effect of garlic (Allium sativum) on Staphyloccus aureus: An in vitro study. African J. Biotech. 10 (4): 666-669.

Eugen JV, Katrin B and Eckhard N.2005. Antidiabetic Effect of Cinnamomum cassia and Cinnamomum zeylanicum In vivo and In vitro. Phytother. Res. 19: 203–206.

Egwaikhinde PA, Okeniyi SO, Akporhonor EE and Emua SO. 2008. Studies on bioactive metabolites constituents and antimicrobial evaluation of leaf extracts of Eucalyptus globule. Agri. J. 3(1): 42-45.

Elumalaii EK, Chandrasekaran N, Thirumalaii T, Sivakumari C, Viviyan TS and David E. 2009. Achyranthes aspera leaf extract s inhibited fungal groth. Inter. J. PharmTech Res. 1(4):1576-1579.

Fabian D, Sabol M, Domaracka K, Bujnakova D. 2006. Essential oils-their antimicrobial activity against Escherichia coli and effect on intestinal cell viability. Toxicol In vitro 20: 1435-1445.

Fariba B, Hassan R and Homeyra E. 2010. In vitro study of the effects of henna extracts (Lawsonia inermis) on Malassezia species. Jundishapur J Microbiol. 2010(3): 125 – 8.

Fabricant DS and Farnsworth NR. 2001. The value of plants used in traditional medicine for drug discovery. Environ Health Perspect. 109 (Suppl 1): 69 – 75.

Ghani A. 1998.Medicinal Plants of Bangladesh- Chemical Constituents and Uses. Asiatic Society of Bangladesh, Dhaka. 215- 216.

Graul T, Cain AM and Karpa KD. 2009.Lactobacillus and bifidobacteria combinations: a strategy to reduce hospital-acquired Clostridium difficile diarrhea incidence and mortality. Med Hypotheses. 73(2): 194-8.

Geissman TA. 1963. Flavonoid, Compound, Tannis, Lignins, and compounds. M. Florkin, E.H. Stotz (Eds.), Pyrrole Pigments, Isoprenoid Compounds and Phenolic Plants Constitutes, 9, Elsevier, New York.

Gu L, Kelm MA, Hammerstone JF, Beecher G, Holden J, Haytowitz D, Gebhardt S, Prior RL. 2004. Concentrations of proanthocyanidins in common foods and estimations of normal consumption. J Nutr 134: 613-617.

Ghosh A, Das BK, Chatterjee SK and Chandra G. 2008. Antibacterial potentiality and phytochemical analysis of mature leaves of Polyalthia longifolia (Magnoliales: Annonaceae). The South Pacific J. Natural Sci., 26: 68-72.

Girach RD and Khan ASA. 1992. Ethnomedicinal uses of Achyranthes aspera leaves in Orissa (India). Int. J Pharmacogn. 30:113-115.

Ghani A.2003. Medicinal Plant of Bangladesh with Chemical Constituents and Uses. 2nded. Asiatic Society of Bangladesh, Dhaka, p.71-72.

Höfling, JF, Anibal, PC, Obando-Pereda GA, Peixoto IAT, Furletti VF, Foglio MA and Gonçalves RB. 2010. Antimicrobial potential of some plant extracts against Candida species. Braz. J. Biol., 70(4), 1065-68.

Hawkey PM. 1998. The origins and molecular basis of antibiotic resistance. BMJ, 317(7159):657-60.

Hiromu S, Akira K, Tamas K, Tamas J and Masakazu H. 2010. Metallo–allixinate complexes with anti-diabeticand anti-metabolic syndrome activities Metallomics 2: 670-682.

Han ST and Un CC. 2003. Cardiac toxicity caused by Achyranthes aspera. Vet Hum Toxicol. 45(4):212-213.

Ibukun A, Tayo A, Toyin A, Tolu O, and Tolu O. 2007. Evaluation of the Antimicrobial Properties of Different Parts of Citrus Aurantifolia (Lime Fruit) as Used Locally. Afr J Tradit Complement Altern Med. 4(2): 185 – 90.

Imparl-Radosevich J, Deas S, Polansky MM. 1998. Regulation of phosphorylase (PTP-1) and insulin receptor kinase by fractions from cinnamon: implications for cinnamon regulation of insulin signalling. Horm Res 50: 177–182.

Jarvill-Taylor KJ, Anderson RA, Graves DJ. 2001. A hydroxychalcone derived from cinnamon functions as a mimetic for insulin in 3T3-L1 adipocytes. J Am Coll Nutr 20: 327– 336.

Jaber MA and Al-Mossawi A. 2007. Susceptibility of some multiple resistant bacteria to garlic extract. African J. Biotech. 6: 771-776.

Jehan B, Muhammad T, Huma A, Amjad I and Mohammad S. 2011. Effect of different solvent extracted sample of Allium sativum (Linn) on bacteria and fungi. African J. Biotech. 10(31): 5910-5915.

Jayaweera, DMA. 1980. Medicinal Plants used in Ceylon Part 2. National Science Council of Sri Lanka. Colombo.

Jayaprakasha GK, Rao J, Sakariah KK. 2002. Chemical composition of volatile oil from Cinnamomum zeylanicum buds. Z Naturforsch (C) 57: 990-99.

Jayaprakasha GK, Kameyama MO, Ono H, Yoshida M, Rao J. 2006. Phenolic constituents in the fruits of Cinnamomum zeylanicum and their antioxidant activity. J Agr Food Chem 54: 1672-1679.

Josephin SB and Selva MT. 2012. Antimicrobial activity of Adhatoda vasica against clinical pathogens. Asian J Plant Sci and Res 2(2): 83-88.

Kapadia SG and Yagnik BN. 2013. Current Trend and Potential for Microbial Biosurfactants. Asian J Experimental Biol Sci, 4, 1-8.

Khopde SM, Priyadarshini KI, Guha SN, Satav JG, Venkatesan P and Rao MNA. 2000. Biosci. Biotech. Biochem. 64:503-509.

Kaufmann SH. 1989. Induction of endonucleolytic DNA cleavage in human acute myelogenous leukemia cells by etoposide, camptothecin, and other cytotoxic anticancer drugs: a cautionary note, Cancer Res., 49: 5870-5878.

Khan A, Safdar M, Khan MMA, Khattak KN, Anderson RA. 2003. Cinnamon improves glucose and lipids of people with type 2 diabetes. Diabetes Care 26: 3215–3218.

Khan A, Bryden NA, Polansky MM, Anderson RA. 1990. Insulin potentiating factor and chromium content of selected foods and spices. Bio Trace Element Res 24: 183–188.

Kar A, Choudhary BK, Bandyopadhyay NG. 2003. Comparative evaluation of hypoglycemic activity of some Indian medicinal plants in alloxan diabetic rats J Ethenopharmacol, 84(1):105-108.

Kamath JV, Rana AC and Chowdhury AR. 2003. Pro-healing effect of Cinnamomum zeylanicum bark. Phytother Res 17: 970-972.

Liersch BA. 1992. *In*: Haensel R, Keller K, Rimpler G, Schneider G (eds) Hagers Handbuch der Pharmazeutischen Praxis, V. Springer-Verlag, Berlin 7, pp 54-59.

Lawless J. 2002. The Encyclopedia of Essential Oils, Thorsons: London.

Lee KG, Shibamoto T. 2002. Determination of antioxidant potential of volatile extracts isolated from various herbs and spices. J Agr Food Chem 50: 4947-4952.

Lee JS, Jeon SM, Park EM, Huh TL, Kwon OS, Lee MK, Choi MS. 2003. Cinnamate supplementation enhances hepatic lipid metabolism andantioxidant tdefense systems in high cholesterol-fed rats. J Med Food 6: 183-191.

Mohinder K, Thakur V and Rana RC 2005. Antimicrobial Properties of Achyranthes aspera. Ancient Sci. Life, XXIV (4): 1-6.

Murray PR, Baron EJ, Pfaller MA, Tenover FC, Yolken RH. 1999. Eds. Manual of clinical microbiology. 7th ed. Washington, DC: ASM.

Manandhar NP. 2002. Plants and People of Nepal Timber Press. Oregon pp. 526-7.

Mukul SA, Uddin MB and Tito MR. 2007. Medicinal plant diversity and local healthcare among the people living in and around a conservation area of northern Bangladesh, Inter J Forest Usufructs Manag, 8(2): 50-63.

Mastura M, Azah MAN, Khozirah S, Mawardi R, Manaf AA. 1999. Anticandidal and antidermatophytic activity of Cinnamomum species essential oils. Cytobios 98: 17-23.

Maurya RAand Jayendra. 2010. Chemistry and pharmacology of Withania coagulans: an Ayurvedic remedy. J Pharm and Pharmaco, 62(2): 153-160.

Nayak BS, Isitor G, Davis EM and Pillai GK. 2007. The evidence based wound healing activity of Lawsonia inermis Linn. Phytother Res. 21(9): 827 – 31.

Nadkarni KM, Nadkarni AK. 1999. Indian Materia Medica- with Ayurvedic, Unani-Tibbi, Siddha, Allopathic, Homeopathic, Naturopathic and Home remedies. Popular Prakashan Private Ltd., Bombay, India. ISBN No. 81-7154-142-9.

Onyeagba RA, Ugbogu OC, Okeke CU and Iroakasi O. 2006. Studies on the antimicrobial effects of garlic (Allium sativum Linn), ginger (Zingiber officinale Roscoe) and lime (Citrus aurantifolia Linn). African J. Biotech. 3: 552-554.

Pan SY, Gao SH, Zhou SF, Tang MK, Yu ZL and Ko KM. 2012. New perspectives on complementary and alternative medicine: an overview and alternative therapy. Alter Ther Health Med. 18: 20 – 36.

Pradhan SJ, Nesler KR, Rosen SF, Kato Y, Nakamura A, Ramaswami M and Barbee SA. 2012. The conserved P body component HPat/Pat1 negatively regulates synaptic terminal growth at the larval Drosophila neuromuscular junction. J. Cell Sci. 125(24): 6105--6116.

Perry LM. 1980. Medicinal Plants of East and Southeast Asia: Attributed Properties and Uses, pp. 149–150. MIT Press, Cambridge.

Pranithanchai W, Karalai C, Ponglimanont C, Subhadhirasakul S and Chantrapromma K. 2009. Cassane diterpenoids from the stem of Caesalpinia pulcherrima. Phytochemistry 70(2): 300-304.

Reuter HD, Koch HP and Lawnson DL. 1996. Therapeutic effects and Applications of garlic and its Preparation. In: Garlic: the science and Therapeutic Applications of Allium sativum L. and related Species, 2nd Edn., Koch, H. P. and Lawnson, D. L. (Eds), pp.135-212.

Ramesh L, Chinnappa RV and Abhay KK. 2011. Potential antibacterial and antifungal activity of achyranthes aspera L. Recent Research in Science and Tech. 3(4): 53-57

Raghu HS and Ravindra P. 2010. Antimicrobial activity and phytochemical study of Phyllanthus Emblica linn. Inter. J. Pharm. Stud. Res., IJPSR I(I) 30-33.

Raffauf RF. 1996. A Guide to Their Discovery and Distribution. Hawkworth Press, Inc. New York. p.35.

Rao YK, Fang SH and Tzeng YM. 2005. Anti-inflammatory activities of flavonoids isolated from Caesalpinia pulcherrima. J. Ethnopharmacology, 100(3): 249 - 253.

Roffey B, Atwal A,Kubow S. 2006. Cinnamon water extracts increase glucose uptake but inhibit adiponectin secretion in 3T3-L1 adipose cells. Mol. Nutr. Food Res. 50: 739-745.

Ranjbar A, Ghasmeinezhad S, Zamani H, Malekirad AA, Baiaty A, Mohammadirad A, Abdollahi M. 2006. Antioxidative stress potential of Cinnamomum zeylanicum in humans: a comparative cross-sectional clinical study. Therapy 3: 113-117.

Selvanayagam ZE, Gnanavendan SG, Balakrishnan K, Rao RB. 1994. Antisnake venombotanicals from Ethnomedicine. J Herbs Spices Med Plants 2:45-100.

Sukanya SL, Sudisha J, Hariprasad P, Niranjana SR, Prakash HS and Fathima SK. 2009. Antimicrobial activity of leaf extracts of Indian medicinal plants against clinical and phytopathogenic bacteria. African Journal of Biotechnology, 8(23): 6677-82.

Srikumar RN, Jeya PEM, Shankar S, Manikandan R, Vijayakumar R, Thangaraj K, Vijayananth R and Sheeladevi UAR. 2007. Evaluation of the growth inhibitory activities of triphala against common bacterial isolates from HIV infected patients. Phytotherapy Research. 21(5): 476-80.

Susumu Y, Shigeo K, Norihiro H, Tsuyoshi U, Hiromichi M and Shizutoshi N. 1987. Antifungal Activity of Ajoene Derived from Garlic. Applied and Environmental Microbiology, 53(3): 615-617.

Schiebinger LL. 2004. Plants and empire: colonial bioprospecting in the Atlantic world. Harvard University Press: Cambridge, Mass, pp.10-17.

Sutar NR, Garai US, Sharma PG and Yadav G. 2011. Pharmacognostic studies of the Achyranthes aspera leaves. Pharmacie Globale (IJCP), 5 (10): 1-3.

Simic A, Sokovic MD, Ristic M, Grujic-Jovanovic S, Vukojevic J and Marin PD. 2004. The chemical composition of some Lauraceae essential oils and their antifungal activities, Phytother. Res. 18: 713-717.

Shobha KL, Bernaitis L, Akshatha SJ, Revathi PS, Ramachandra L. 2014. Antibacterial activity of selected siddha herbo-mineral preparations against clinically isolated enterococcus species. International Journal of Pharmacology and Toxicology, 2 (2): 38-40.

Tahsina SH, Mohammad AH, Mohammad GM, David JB, Masayuki F and Lam-Son P. 2016.Tran. Methylglyoxal: An Emerging Signaling Molecule in Plant Abiotic Stress Responses and Tolerance. Front Plant Sci. 7: 1341.

Teferi G and Hahn HJ. 2002. Treatment of malaria in Ethiopia folk medicine. Trop. Doc. 32: 206-207.

Tang C, Chen G, Chen D, Xin H and Gao L. 2009. Study on Antimicrobial Activity of Crude Extract from Phyllanthus emblica L. Fruits.Chinese J. 30(7): 106-108.

Thakur RS, Puri HS and Husain A. 1989. Major Medicinal Plants of India. Central Institute of Medicinal and Aromatic Plants, Lucknow, India.

Treadway L. 1994. Amla Traditional food and medicine. J. Amer. Bot. Coun. 31: 26.

Tsarong TJ. 1994. Tibetan Medicinal Plants. First edition. Tibetan Medical Publications, India. ISBN No. 81-900489-0-2.

Tahiliani P and Kar A. 2000. Achyranthes aspera elevates thyroid hormone level and decrease hepatic lipid peroxidation in male rats. J. Ethanopharmacol. 71: 527-532.

Torssel KBG. 1993. Natural Product Chemistry. A mechanistic and biosynthetic approach to secondary metabolism. John Wiley & Sons Ltd, London.

Tailang M, Bhaskar KG, Amrish S. 2008. Antidiabetic Activity of Alcoholic Extract of Cinnamomum zeylanicum Leaves in Alloxon Induced Diabetic Rats People's Journal of Scientific Research, 1: 9-11.

Tomaino A, Cimino F, Zimbalatti V, Venuti V, Sulfaro V, De Pusquale A, Saiha A. 2005. Influence of heating on antioxidant activity and the chemical composition of some spice essential oils. Food Chem. 89: 549-554.

Ureghe UBO, Ehwarieme DA and Eboh DO. 2010. Antibacterial activity of garlic and lime on isolates of extracted carious teeth. African J. Biotech.9(21): 3163-3166.

Udupa KN. 2002 Ayurveda for Promotion of Health.Journal of Ayurveda, 3. Williamson EM (editor), Major Herbs of Ayurveda, Churchill-Livingstone, London.

Valsaraj R, Pushpangadan P, Smitt U W, Andersen A and Nyman U. 1997. Antimicrobial screening of selected medicinal plants from India. J Ethnopharmacol. 58:75-83.

Van FR. 1982. Current view on the mononuclear phagocyte system. Immunobiol. 161:178–185.

Verspohl EJ, Katrin B and Neddermann E. 2005. Anti-diabetic effect of Cinnamomum cassia and Cinnamomum zeylanicum in vivo and in vitro. Phytotherapy Research, 19(3): 203 -206.

Wanner J, Schmidt E, Bail S, Jirovetz L, Buchbauer G, Gochev V, Girova T, Atanasova T and Stoyanova A. 2010. Chemical composition, olfactory evaluation and antimicrobial activity of selected essential oils and absolutes from Morocco. Nat Prod
Commun. 5(9):1349-54.

Winslow LC and Kroll DJ. 1998.Herbs as medicines. Arch Intern Med. 158(20): 2192-9

Xiaoli L, Mouming Z, Jinshui W and Wei L. 2009. Antimicrobial and antioxidant activity of emblica extracts obtained by supercritical carbon dioxide extraction and methanol extraction. J. Food Biochem. 33: 307–330.

Yehouenou B, Ahoussi E, Sessou P, Toukourou F and Sohounhloue D.2012. Chemical composition and antimicrobial activities of essential oils (EO) extracted from leaves of Lippia rugosa A. Chev against foods pathogenic and adulterated microorganisms. Afr. J. Microbiol. Res. 6: 5496-5505.

2.2 MEDICINAL VALUE AND ANTIMICROBIAL ACTIVITY OF SOME SELECTED PLANTS

SOURCE:

1. *Zakaria Ahmed. 2013. Medicinal and Antimicrobial Properties of Mushrooms. A scientific journal of Comsats – Science Vision. 19(1 & 2): 41-48.*
2. *Sajib Chakrabortty and Zakaria Ahmed. 2013. Study of antimicrobial activity of Ayurvedic and Unani medicine and their comparative analysis with commercial antibiotics. IMPACT: international Journal of Research in Applied, Natural and Social Sciences (Impact: IJRANSS). 1(3): 63-74.*
3. *Zakaria Ahmed, Md. Mehedi Hasan Magnet and Mesbah Uddin Md. Rezwanur Nayem. 2014. Medicinal Value and Antimicrobial Activity of Adhatoda Vasica and Cinnamomum Zeylanicum - A Mini Review. Journal of Medical and Pharmaceutical Innovation. 1 (6): 24-29.*
4. *Zakaria Ahmed, Taslima Rahman and Sanchita Sarkar. 2019. Antimicrobial analysis of some selected Ayurvedic and Unani medicines. International Journal of Scientific Research and Engineering Development. 2(6): 561-569.*
5. *Md. Mahabub Ali, Balaram Ray, Begum Rokeya, Mst. Aleya Nasreen and Zakaria Ahmed. 2019. Potential Healing Powers with Jute Plant- A Review. International Journal of Sciences: Basic and Applied Research (IJSBAR). 48(5): 10-23.*
6. *Aleya Nasreen, Zakaria Ahmed, Mahbub Ali and Tahmina Shimu. 2021. Jute leaves: A potential sources of lycopene, International Journal of Vegetable Science, DOI:10.1080/19315260.2021.1931619.*
7. *Zakaria Ahmed and Shuranjan Sarkar. 2022. Review on jute leaf: A powerful biological tool. International Journal of Scientific Research Updates. 4(1): 64–85.*
8. *Taslima Rahman, Zakaria Ahmed and Shuranjan Sarkar. 2022. Analysis of the antibacterial properties of jute leaf. Open Access Research Journal of Life Sciences. 4(1): 51–55.*
9. *Mst. Aleya Nasreen, Zakaria Ahmed, Md. Mahabub Ali and Tahmina. 2022. Determination of β-carotene in jute leaves by spectrophotometry and thin layer chromatography. World Journal of Biology Pharmacy and Health Sciences. 9(2): 11–20.*

2.2.1 MUSHROOMS [Zakaria 2013]

Mushrooms are unique within the fungal kingdombecause they produce complex fruiting body. They aresaprophytes. All of the mushrooms are placed in adivision Eumycota meaning 'The True Fungi. They aredivided into other groups depending on the structure of their fruiting bodies and various other macro- and microscopic characteristics. The two major groups of Eumycota are Basidiomycetes and Ascomycetes. They include members of the Basidiomycota, within the class of Homobasidiomcetae. Mushrooms have been a food supplement in various cultures and are cultivated and eaten for their edibility and delicacy. They fall between the best vegetables and animal protein source. Mushrooms are considered as source of proteins, vitamins, fats, carbohydrates, amino acids and minerals (**Jiskani 2001**). All essential amino acids as well as water-soluble vitamins and all the essential minerals (**Buigut 2002**). Mushrooms are good sources of vitamins like riboflavin, biotin and thiamine. Ogundana and Fagade (**Ogundana and Fagade 1981**) indicated that mushroom is about 16.5% dry matter out of which 7.4% is crude fiber, 14.6% is crude protein and 4.48% is fat and oil. Protein contents vary between 4 to 9% in Auricularia species and between 24 to 44 % in Agaricus species. The protein value of mushrooms is twice as that of asparagus and potatoes, four times as that of tomatoes and carrots, and six times as that of oranges (**Jiskani 2001**). Their energy value also varies according tospecies, which is nearly equal to that of an apple. Inrecent times, mushrooms have assumed greater importance in the diets of both rural and urban dwellers. They are also relatively much cheaper than beef, pork and chicken that contain similar nutrients. Many mushrooms have long been valued as tasty,nutritious food by different societies worldwide. Mushrooms are popular and valuable functional food, low in calories and high in minerals, essential amino acids, vitamins and fibers.The results of proximate analysisby Adejumo andAwosanyaof four edible species of mushroom collected from Nigeria (**Adejumo and Awosanya 2005**) indicated that*Termitomycesmammiformis* was a very good source of crudeprotein, crude fiber, ash, calcium and manganese. *Russula vesca* was the richest in carbohydrate and magnesium, while *Lactarius triviralis* was found to be the richest in moisture content, iron and copper. It is also a good source of carbohydrate, calcium andmanganese.*Lentinustigrin*is the richest in dry matter, and is also rich in carbohydrate, magnesium and copper. It was observed that lipids, sodium and phosphorus contents of the four species were generallyverylow.Theresultsofmineralvalues(**Adejumo and Awosanya 2005**)of the four edible species of mushrooms clearlyindicate the potential for their use as sources of good quality food. It was observed that these four edible mushrooms hold tremendous promise in complementingtheproteinandmineralsupplydeficits prevalentindevelopingcountries.

2.2.1.1 ANTIMICROBIAL PROPERTIES Different types of mushrooms have different antimicrobial properties that are useful for human health. *Pleurotus tuber-regium*is useful in some combinations to cure headache, stomach ailments, cold andfever, asthma, smallpox and high blood pressure (**Oso 1977, 1977a, 1977b**), while*Lentinustuber regium*and *L. tigrinus*are used for treating dysentery and blood cleansing, respectively. *Auricularia* species have been traditionally used for treating hemorrhoids and various stomach ailments (**Chang 1996**). *Boletus edulis* and *Lactarius* species are also used for killing flies, while the puffballs are used for healing wounds (**Harkonen 1998, Delena 1999**). They are also recommended to diabetic and anemic persons, owing to their low carbohydrate and high folic acid content. Some mushrooms are known for possessing anti-allergic, anti-cholesterol, anti-tumor and anti-cancer prorates (**Jiskani 2001**). It was observed that possess some curative effects against some bacterial infections and intestinal disorders (**Jonathan 2002,Oso 1977a, 1977b**). Jonathan et al(**2002**) also reported the antagonistic effect of extracts of some Nigerian higher fungi against selected pathogenic microorganisms. Olawuyi et al(**2011**) reported the antibacterial activities of the distilled water, ethanolic and chloroform extracts of against three Gram-negativebacteria (*E. coli, Proteus mirabilis* and *Pseudomonas aerogenosa*) and two Gram-positive bacteria (*Bacillus cereus* and*Staphylococcus aureus*) where the antibacterial activities of the tested mushroom vary and are target microbes specific. The activity of extract of chloroform and ethanol against was significantly higher. According to Olawuyi et al(**2011**), Gram-negative bacteria are more sensitive to fresh *F. lignosus* than Gram-positive bacteria.

2.2.1.2 MEDICINAL USES Recently, the demand for more effective and safertherapeutic agents for the chemoprevention of human cancer has increased. Natural products produced by plants and their synthetic derivatives are expected to play an important role in the development of innovative agents to inhibit the onset of cancer (**Taji et al 2007**). Macro fungi, such as mushrooms and enthomopathogenic fungi, are good sources of natural medicines that exert antitumor activity. Polysaccharides are the best known and most potent mushroom-derived substances because they inhibit the growth of many types of tumors (**Wasser 2002, Moradali et al 2007**). Natural antitumor isolated from mushrooms include, acidic and neutral compounds with different types of glycosidic linkages, as well as some that are bound to protein or peptide residues such as polysaccharide protein complexes (**Wasser 2002, Moradali et al 2007, Chen and Chang 2004**). Animal studies show mushrooms have anti-inflammatory activity. Inotilone, quercinol, antcin A, and benzocamphorin F, are anti-inflammatory compounds isolated from mushrooms. *Pholiota squarrosa*contains xanthine oxidase inhibitors. *Coprinopsis atramentaria*contains coprine, which metabolizes to 1-aminocyclopropanol, an inhibitor of acetaldehyde dehydrogenase. A number of mushrooms inhibit 5-reductase and/or aromatase invitro. In Southeast Asia, especially in China and Japan, mushrooms have long been acknowledged for their medicinal and analeptic qualities in addition to their desirable flavors and nutritional value. Several mushroom species display anti-oxidant power (**Benzie and Strain 1996, Benzie et al 1998**),and mushroom-derived polysaccharoproteins arereported to scavenge active oxygen species (**Liu et al 1997**).The medicinal properties of mushrooms have longbeen recognized, especially in Oriental cultures, and modern techniques have identified numerous bioactive mushroom components, which are variously reported to exhibit anti-cancer, anti-tumor, anti-viral, immunomodulatory, hypocholesterolaemic and hepatoprotective activities (**Chang 1996**). More recently, it has been shown that fruit bodies of and contained bioactive compounds that prevented HO-induced oxidative damage to cellular DNA(**Shi et al 2002**).Although the medicinal effects of products are well documented (**Chang and Buswell 1999**), there are few reports attributing medicinal properties to even though it is the most widely cultivated and consumed edible mushroom (**Chang 1999**). Quinoid compounds obtained from this mushroom have been reported to suppress the propagation of mouse ascites tumor (**Graham et al 1977**), and a lectin from this species also reversibly inhibited the proliferation of human colon carcinoma cells (**Yu et al 1993**). Both cellular components and secondary metabolites of a large number of mushrooms have been shown to affect the immune system of the host, and therefore could be used to treat diseases of clinical importance (**Oso 1977**).

It is well known that mushroom polysaccharides primarily exert their antitumor activity via activation ofthe immune response of the host organism (immunoenhancing activity) (**Kim et al 2006**). This indicates that mushroom polysaccharides do not directly kill tumor cells. Rather, they help the host to adapt to various biological stresses and exert a nonspecific action on the host, supporting some or all of the major systems. Because mushroom polysaccharides cause no harm and place no additional stress on the body, they are regarded as biological response modifiers. Immunoenhancing activity has been observed in many mushroom polysaccharides (**Zhang et al 2007**).*Ganoderma lucidum* has been widely used in China and Japan for thousands of years for the treatment of various diseases, including cancers, confirming its possibility of revitalization and curing of different illnesses. It acts as antitumour, anti-inflammatory, antiviral (e.g. anti-HIV), antibacterial, antiparasitic, immunomodulating and hepato protective. It has a role in blood pressure regulation, against cardiovascular disorders and chronic bronchitis, like kidney tonic and nerve tonic (**Wasser and Weis 1999a**). Pharmacologically, a number of the water-soluble polysaccharides have demonstrated antitumor and immunostimulating activities. *Karst* is one of the most often used mushrooms in traditional medicine of Far Eastern people. Because of its bitter taste and wooden built it is not suitable for nutrition, but the bioactive substances extracted from this mushroom possess very important medicinal characteristics.

Polysaccharides, β-glucans, obtained through theprocesses of hot water extraction, alcoholprecipitation and dialyses refining had influence on thereduction of B cells (a human B cell line transformed byEpstein-Barr virus) and T cells (an immortalized line ofhuman T lymphocyte cells). Among all testedconcentrations of mushroom extract, the mostintensive influence showed concentration of onemg/ml, which reduced the number of B cells by61.46%, while in the case of T cells their number werereduced by 57.14% (**Mizuno et al 1995**). Edible mushrooms havebeneficial effects on health and in the treatment ofdisease through their immunomodulatory, anti-neoplastic and lipid-reducing properties (**Wasser and Weis 1999b, Anita et al 2009**). TheShiitake mushroom (*Lentinula edodes*), for example,has served as a model for investigating functionalmushrooms and isolating pure compounds for pharmaceutical use (**Jong and Birmingham 1993**). Water extracts of theShiitake fruiting bodies have been shown to preventtumor growth in mice (**Chang 1996,Mitamura et al 2000, Ng and Yap 2002**). Mushroom'spolysaccharides, especially the high-molecular-weight b-D-glucan have been considered to have anti-cancer activity (**Cheung et al 2002**). Iris et al (**2006**) stated thathas anti-proliferative and pro-apoptoticactivities where they described a newly identified low-molecular-weight α-glucan with promising anti-tumorigenic properties and demonstrated its directeffect on colon cancer cell proliferation via induction ofprogrammed cell death.A major strategy for colon cancer chemoprevention isthe search for nutritional components directed atinducing apoptosis of cancer cells. Edible mushroomshave beneficial effects on health and in the treatmentof disease through their immunomodulatory, anti-neoplastic, and lipid-reducing properties (**Kidd 2000, Lee et al 2003, Zusman et al 1997**) has been used as a folk medicinedue to its antitumor properties against several types ofcancer cells (**Taji et al 2007,Kahlos et al 1986, Saar 1991**). In some cases, thetherapeutic efficiency of has beendemonstrated clinically. For example, a decoction offungal sclerotia did not show toxic effects and hasbeen used in the treatment of cancers and digestivesystem diseases (**Park et al. 2004,Nakata et al 2007**). Recently, in addition topolysaccharides, many polyphenolic compoundssuch as triterpenoids, steroids and ergosterolperoxides from have been found to possessbiological activities, including antioxidant,antibacterial, hepatoprotective, platelet aggregationinhibitory and antitumor effects (**Kahlos et al 1986,Park et al 2005,Shin et al 2001, Cui et al 2005, Hyun et al 2006, Mazurkiewicz 2006, Lee et al 2007, Nakajima et al 2007, Shi et al 2008, Youn et al 2008**). Sung et al (**2009**) reported that the hot water extract ofexerts inhibitory activity against theproliferation of human colon cancer cells (HT-29) andtheir results suggested that this mushroom would beuseful as an antitumor agent via the induction ofapoptosis and inhibition of the growth of cancer cellsthrough up-regulation of the expression ofproapoptotic proteins and down-regulation ofantiapoptotic proteins (**Sung et al 2009**). Mizuno reported thatpolysaccharides from fungal sclerotia, which areknown to include hetero-polysaccharide andhomoglycan, showed strong antitumor effects, whilepolysaccharides from cultured mycelia did not. However, Kim et al (**2005, 2006**) reported that thepolysaccharides extracted from cultivated mycelia(termed endo-polysaccharides) also have anticanceractivity. *Agaricus subrufescens*extracts demonstratedimmunomodulatory activities (**Ren et al 2008, Wu et al 2007**) andactivity against several cancer cell lines (**Yu et al 1993**). *Agaricussubrufescens* (*Agaricus blazei/brasiliensis/sylvaticus*)is another medicinal mushroom associated with Braziland Japan. There was an evidence for usingextracts for certain cancers (**Fortes et al 2009, Ahn et al 2004, Hetland et al 2008)** whereas has anticancer activity in vivo(**Ji et al 2012**).

Auricularia auricular extracts demonstratedhypoglycemic in vivo and anticancer, anticoagulant,and anticholesterol activities in vitro (**Takeujchi et al 2004, Misaki et al 1981**). An*Auricularia polytricha* isolate inhibited sarcoma in vivo (**Song and Du 2010**). A *Boletus edulis*isolate inhibited several cancercell lines and was found to bind a cell-specific T-antigen disaccharide (**Bovi et al 2011**). A *Coprinus comatus* extractinhibited adenocarcinoma (**Dotan et al 2010**).*Dictyophora indusiata*creates dictyophorines, dictyoquinazols, and tyrosinase inhibitors (**Ker et al 2011**). Epidemiological studiesin Japan have associated*Flammulina velutipes*consumption with lower cancer rates (**Monro 2003**). Proflamin,an isolate of *Flammulina velutipes,* has anticanceractivity in vivo (**Ikekawa et al 1985**). *FIP*-5 is an immunomodulatoryprotein isolated from *Flammulina velutipes.* Many*Grifola frondosa*studies indicate potential anticancerand antihyperglycemic activities. Antiatherosclerotic,antifungal and immunomodulatory compounds havebeen isolated from*hypsizygus tessellatus.*Lentinan is an isolate of *Lentinula edodes*mycelia. InJapan and China, lentinan is an adjuvant for gastriccancer (**Ina and Ando 2012**). AHCC is an extract derived fromfermented *Lentinula edodes*mycelia. Theisolate eritadenine, is an inhibitor of S-adenosyl-L-homocysteine hydrolase, and hashypocholesterolemic activity.*Lignosus rhinocerus* is amedicinal mushroom associated with Malaysia (**Lee et al 2011**). A*Morchellaesculenta* isolate demonstratedimmunomodulatory activity*in vitro*(**Duncan et al 2002**).*Phallus mpudicus* extracts have been clinically researched inrelation to venous thrombosis (**Kuznecova et al 2007**).*Pholiota nameko*creates compounds with antiinflammatory,immunomodulatory, and hypolipidemic activities (**Li et al. 2008, Li et al 2012, Li et al 2010**). A *Pleurotus eryngii* isolate demonstratedanticancer activity*in vitro* (**Wu et al 2010**).*Pleurotus eryngii*extracts have immunomodulatory activities*in vitro*(**Nozaki et al 2008**).

A number of studies indicate*Pleurotus ostreatus*consumption lowers cholesterol levels *in vivo.*Research with *Pleurotus ostreatus* extracts indicatespotential anticancer and immunomodulatory activities (**Zusman et al 1997,Sliva 1992, Bobek and Galbavy 2001**). *Saparassis crispa*has anticancer andimmunomodulating activity *in vivo*(**Ohno et al 2000, Harada et al 2002a, Harada et al 2002b, Nameda et al. 2003, Kim et al 2010**).A *Tremella fuciformis*isolate protected against effectsof radiation *in vivo. Tremella mesenterica*has potentialanticancer and immunomodulating activities (**Vinogradov et al 2004, Vinogradov et al 2004b, Elisashvili 2002**). *Tricholoma matsutake*isolates may haveanticancer and immunomodulating activities (**Ishihara et al 2003, Ishihara et al 2004, Ebina 2003, Hoshi et al 2005, Hoshi et al 2008, Byeon et al 2009, Ebina 2005**).*Volvariella volvacea,* a mushroom associated withThailand and Vietnam, has anticancer activity *in vitro*(**Wu et al 2010**).

2.2.1.3 CONCLUSION

It has been known for many years that selectedmushrooms of higher Basidiomycetes origin areeffective against cancer. There are approximately 650species of higher Basidiomycetes that have beenfound to possess antitumor activity. Searching for newantitumor and other medicinal substances frommushrooms and studying the medicinal value of thesemushrooms has become a matter of great significance. Some countries classify mushroomisolates as antineoplastic compounds. Mushroomsand other forms of fungi create antimicrobial andantiviral compounds. Now-a-days consumption ofmedicinal mushrooms is increasing strongly.Therefore, there is a room for improvement ȯftechnological knowledge in the determination ofantimicrobial and medicinal substances from selectedmushrooms, which will be useful in production ofvaluable antibiotic drugs in pharmaceutical industries.

2.2.1.4 REFERENCES

Adejumo TO and Awosanya OB. 2005. Proximate and mineral composition of four edible mushroom species from South Western Nigeria. Afr. J. Biotech., 4(10): 1084-1088.

Anita SK, Maja SK and Miomir PN. 2009. Influence of bioactive compounds extracted from mushroom Ganoderma lucidum on B and T cells. Proc. Nat. Sci, Matica Srpska Novi Sad, 116: 217-223.

Ahn WS, Kim DJ, Chae G. et al., 2004. Natural killer cell activity and quality of life were improved by consumption of a mushroom extract, Agaricus blazei Murill Kyowa, in gynecological cancer patients undergoing chemotherapy. Inter. J. Gynecological Cancer. 14(4): 589-94.

Bobek P and Galbavy S. 2001. Effect of pleuran (beta-glucan from Pleurotus ostreatus) on the antioxidant status of the organism and ondimethylhydrazine-induced precancerous lesions in rat colon. British J. Biomedical Sci. 58(3): 164-8.

Buigut SK. 2002. Mushroom production in sustainable small-scale farming system-opportunities and constraints: a survey of Uasin Gishu district. In: Proceedings of the Holticulture seminar on Sustainable Horticultural Production in the Tropics at Jomo Kenyatta University of Agriculture & Technology, Juja, Kenya 3 -6 October, 2001. Eds. Wesonga JM, Losenge T, Ndungu CK, Fricke A.

Benzie IFF and Strain JJ. 1996. The reducing ability of plasma as a measure of antioxidant power- the FRAP assay. Anal. Biochem. 239: 70-76.

Benzie IFF, Wu I and Buswell JA. 1998. Antioxidant activity of mushrooms and mushroom products. In: Abstracts 2nd Intl. Conf. Natural Antioxidants and Anticarcinogens in Nutrition, Health and Disease. Helsinki, Finland.

Bovi M, Carrizo ME, Capaldi S. et al. 2011. Structure of a lectin with antitumoral properties inking bolete (Boletus edulis) mushrooms. Glycobiology. 21(8): 1000-9.

Byeon SE, Lee J, Lee E. et al. 2009. Functional activation of macrophages, monocytes and splenic lymphocytes by polysaccharide fraction from Tricholoma matsutake. Archives of Pharmacal Research. 32(11): 1565-72.

Chang R. 1996. Functional properties of edible mushrooms, Nutr. Rev. 54: S91-S93.

Chen YY and Chang HM. 2004. Antiproliferative and differentiating effects of polysaccharide fraction from fu-ling (Poria cocos) on human leukemic U937 and HL-60 cells. Food Chem Toxicol., 42: 759-769.

Chang ST and Buswell JA. 1999. Ganodermalucidum (Curt.: Fr.) P. Karst. (Aphyllophoromycetideae)- A mushrooming medicinal mushroom. Intl. J. Med. Mushrooms 1: 139-146.

Chang ST. 1999. World production of cultivated edible and medicinal mushrooms in 1997 with emphasis on Lentinus edodes, Berk. Sing. In China. Intl. J. Med. Mushrooms, 1: 291-300.

Cheung NK, Modak S, Vickers A, and Knuckles B. 2002. Orally administered b-glucansenhance anti-tumor effects of monoclonalantibodies, Cancer Immunol. Immunother. 51(10): 557-564.

Cui Y, Kim DS and Park KC. 2005. Antioxidant effect of Inonotus obliquus. JEthnopharmacol. 96: 79-85.

Delena T. 1999. Edible and Useful Plants of Texas and the South West- A Practical Guide. 2003-5 University of Texas Press. p. 542.

Dotan N, Wasser SP and Mahajna J. 2010. The Culinary-Medicinal Mushroom Coprinus comatus as a Natural Anti-androgenic Modulator. Integrative Cancer Therapies. 10(2): 148-59.

Duncan CJ, Pugh G, Nirmal P. et al. 2002. Isolation of a Galactomannan That Enhances Macrophage Activation from the Edible Fungus Morchella esculenta. J. Agri. Food Chemistry. 50(20): 5683-5.

Ebina T. 2005. Antitumor effects of intratumoral injection of Basidiomycetes preparations (in Japanese). Gan to kagaku ryoho. 32 (11): 1654-6.

Elisashvili V, Wasser SP and Tan KK. 2002. Hypoglycemic, Interferonogenous, and Immunomodulatory Activity of Tremellastin from the Submerged Culture of Tremella mesenterica Retz.: Fr. (Heterobasidiomycetes), Inter. J. Medicinal Mushrooms. 4 (3): 121-123.

Ebina T. 2003. Activation of antitumor immunityby intratumor injection of biological preparations. (in Japanese). Gan to kagaku ryoho, 30(11): 1555-8.

Fortes RC, Novaes MRC, Recova G and Melo AL. 2009. Immunological, hematological and glycemia effects of dietary supplementation with Agaricus sylvaticus on patients' colorectal cancer, Experimental Biology and Medicine. 234(1): 53-62.

Graham DG, Tye RW and Vogel FS. 1977. Inhibition of DNA polymerase from L1210 murine leukemia by a sulfhydryl reagent from Agaricus bisporus. Cancer Res. 37: 436-9.

Harkonen M. 1998. Uses of mushrooms by Finnsand Karelians. Int. J. Circumpolar Health, 57(1): 40-55.

Hoshi H, Yagi Y, Iijima H. et al. 2005. Isolation and Characterization of a Novel Immunomodulatory a-Glucan-Protein Complex from the Mycelium of Tricholoma matsutakein Basidiomycetes. J. Agri. Food Chemistry 53 (23): 8948-56.

Hoshi H, Iijima H, Ishihara Y. et al. 2008. Absorption and Tissue Distribution of an Immunomodulatory a-d-Glucan after Oral Administration of Tricholoma matsutake. J. Agri. Food Chemistry. 56 (17): 7715-20.

Hyun KW, Jeong SC, Lee DH, et al. 2006. Isolation and characterization of a novel platelet taggregation inhibitory peptide from the medicinal mushroom. Inonotus obliquus. Peptides 27: 1173-1178.

Hetland G, Johnson E, Lyberg T. et al. 2008. Effects of the Medicinal Mushroom Agaricusblazei Murill on Immunity, Infection and Cancer. Scandinavian J. Immunology. 68(4): 363-70.

Harada T, Miura N, Adachi Y. et al. 2002a. Effect of SCG, 1,3-b-D-Glucan from Sparassis crispa on the Hematopoietic Response in Cyclophosphamide Induced Leukopenic Mice. Biological & Pharmaceutical Bulletin. 25 (7): 931-39.

Harada T, Miura NN, Adachi Y. et al. 2002b. IFN-gInduction by SCG, 1,3-b-D-Glucan from Sparassis crispa, in DBA/2 MiceIn Vitro. J. Interferon & Cytokine Research. 22(12): 1227-39.

Iris L, Dana F, Shimona G. et al. 2006. Anaqueous polysaccharide extract from the edible mushroom Pleurotus ostreatus induces anti-proliferative and pro-apoptotic effects on HT-29colon cancer cells, Cancer Letters, 244: 61-70.

Ikekawa T, Maruyama H, Miyano T, et al. 1985. Proflamin, a new antitumor agent: preparation, physicochemical properties and antitumor activity. Japanese J. cancer research. 76(2): 142-8.

Ina K. and Ando T. 2012. The use of Lentinan for Treating Gastric Cancer. Anticancer Agents MedChem. PMID 23092289.

Ishihara Y, Iijima H, Yagi Y and Matsunaga K. 2003. Enhanced Recovery of NK Cell Activity inMice under Restraint Stress by the Administration of a Biological Response Modifier Derived from the Mycelia of the Basidiomycete Tricholo mamatsutake. Stress. 6(2): 141-8.

Ishihara Y, Iijima H, Yagi Y. et al. 2004. Inhibition of Decrease in Natural Killer Cell Activity in Repeatedly Restraint-Stressed Mice by a Biological Response Modifier Derived from Cultured Mycelia of the Basidiomycete Tricholoma matsutake. Neuro immunomodulation. 11 (1): 41-8.

Jiskani MM. 2001. Energy potential of mushrooms. The DAWN Economic and Business Review, 15-21, IV.

Ji Y, Zheng MF, Ye SG. et al. 2012. Agrocybeaegerita polysaccharide combined with chemotherapy improves tumor necrosis factor-a and interferon-g levels in rat esophageal carcinoma. Dis Esophagus. doi:10.1111/j.1442-2050.2012.01397.x. PMID 22947091.

Jonathan SG. 2002. Vegetative growth requirements and antimicrobial activities of some higher fungi in Nigeria. Ph.D thesis. University of Ibadan, Nigeria.

Jong SC and Birmingham JM. 1993. Medicinal and therapeutic value of the shiitake mushroom, Adv. Appl. Microbiol. 39: 153-184.

Kim YO, Park HW, Kim JH et al. 2006. Anti-cancer effect and structural characterization of endopolysaccharide from cultivated mycelia of Inonotus obliquus. Life Sci., 79: 72-80.

Kidd PM. 2000. The use of mushroom glucans and proteoglycans in cancer treatment, Altern. Med. Rev. 5 (1): 4-27.

Kahlos K, Kanga L and Hiltunen R. 1986. Anti-tumor activity of triterpenes in Inonotusobliquus. Planta Med, 52: 554.

Kim YO, Han SB and Lee HW. 2005. Immuno-stimulating effect of the endo-polysaccharide produced by submerged culture of Inonotus obliquus. Life Sci., 77: 2438-2456.

Ker YB, Chen KC, Peng CC. et al. 2011. Structural Characteristics and Antioxidative Capability of the Soluble Polysaccharides Presentin Dictyophora indusiata (Vent. Ex Pers.) Fish Phallaceae. Evid Based Complement Alternat Med. 396013. doi:10.1093/ecam/neq041. PMC3136395. PMID 21799678.

Kuznecova G, Jegina K, Kuznecovs S and Kuznecovs I. 2007. Phallus impudicus in thromboprophylaxis in breast cancer patients undergoing chemotherapy and hormonal treatment. The Breast. 16 (S1): S56.

Kim HS, Kim JY, Ryu HS. et al. 2010. Induction of dendritic cell maturation by b-glucan isolated from Sparassis crispa. Internationa lImmunopharmacology. 10(10): 1284-94.

Liu F, Ooi VEC and Chang ST. 1997. Free radical scavenging activities of mushroom polysaccharide extracts. Life Sci. 60: 763-771.

Lee YL, Kim HJ, Lee MS. et al. 2003. Oral administration of agaricus blazei, H1 strain) inhibited tumor growth in a sarcoma 180 inoculation model, Exp. Anim. 52(5): 371-375.

Lee IK, Kim YS, Jang YW. et al. 2007. New antioxidant polyphenols from the medicinal mushroom Inonotus obliquus. Bioorg Med Chem Lett. 17: 6678-6681.

Li H, Zhang M and Ma G. 2010. Hypolipidemic effect of the polysaccharide from Pholiotanameko. Nutrition. 26 (5): 556-62.

Li H, Lu X, Zhang S. et al. 2008. Anti-inflammatory activity of polysaccharide from Pholiota nameko. Biochemistry (Mosc). 73(6): 669-75.

Li H, Liu X, Li Y. et al. 2012. Effects of the polysaccharide from Pholiota nameko on humancytokine network in serum. Int J Biol Macromol 50(1): 164-70.

Lee SS, Tan NH, Fung SY. et al. 2011. Evaluation of the sub-acute toxicity of the sclerotium of Lignosus rhinocerus (Cooke), the Tiger Milk mushroom. J Ethnopharmacol. 138(1): 192-200.

Mizuno T, Saito H, Nishitoba T and Kawagishi H. 1995. Antitumor-active substances from mushrooms, Food Rev. Int. 11: 23-61.

Mitamura T, Sakamoto S, Suzuki S, et al. 2000. Effects of lentinan on colorectalcarcinogenesis in mice with ulcerative colitis, Oncol. Rep. 7(3): 599-601.

Mazurkiewicz W. 2006. Analysis of aqueous extract of Inonotus obliquus. Acta Pol Pharm. 63: 497-501.

Mizuno T, Ando M and Sugie R. 1992. Antitumor activity of some polysaccharides isolated from an edible mushroom, Ningyotake, the fruiting body and the cultured mycelium of Polyporous confluens. Biosci Biotechnol Biochem. 56: 34-41.

Mizuno T, Zhuang C and Abe K. 1999. Antitumor and hypoglycemic activities of polysaccharides from the sclerotia and mycelia ofI no not us obliquus (Pers.: Fr.) Pil. (Aphyllo phoromycetideae). Int J. Med. Mushr.1: 301-316.

Misaki A, Kakuta M, Sasaki T. et al. 1981. Studies on interrelation of structure and antitumor effects of polysaccharides: Antitumor action of periodate-modified, branched (1 goes to 3)-beta-D-glucan of Auricularia auricula-judae, and other polysaccharides containing (1 goes to 3)-glycosidic linkages. Carbohydrate research. 92(1): 115-29.

Moradali MF, Mostafavi H, Ghods S and Hedjaroude GA. 2007. Immunomodulating and anticancer agents in the realm of macromycetes fungi (macrofungi). Int Immunopharmacol., 7: 701-724.

Monro J. 2003. Treatment of Cancer with Mushroom Products. Archives of Environmental Health: An Inter J. 58(8): 533-7.

Ng ML and Yap AT. 2002. Inhibition of human colon carcinoma development by lentinan from shiitake mushrooms (Lentinus edodes), J. Altern. Complement. Med. 8 (5): 581-589.

Nameda S, Harada T, Miura NN. et al. 2003. Enhanced Cytokine Synthesis of Leukocytes by ablucan Preparation, SCG, Extracted from a Medicinal Mushroom, Sparassis crispa. Immunopharmacology and Immunotoxicology 25(3): 321-35.

Nakata T, Yamada T and Taji S. 2007. Structure determination of inonotsuoxides A and Band in vivo anti-tumor promoting activity of inotodiol from the sclerotia of Inonotus obliquus. Bioorg. Med. Chem. 15: 257-264.

Nakajima Y, Sato Y and Konishi T. 2007. Antioxidant small phenolic ingredients in Inonotus obliquus (persoon) Pilat (Chaga). Chem. Pharm. Bull. (Tokyo). 55: 1222-1226.

Ogundana SK and Fagade O. 1981. The nutritive value of some Nigerian edible mushrooms. In: Mushroom Science XI, Proceedings of the Eleventh International Scientific Congress on the Cultivation of Edible Fungi, Australia. pp. 123-131.

Ohno N, Miura NN, Nakajima M and Yadomae T. 2000. Antitumor 1,3-beta-glucanfrom cultured fruit body of Sparassis crispa. Biological & Pharmaceutical Bulletin. 23(7): 866-72.

Oso BA. 1977. Pleurotus tuber-regium from Nigeria. Mycologia, 69: 271-279.

Oso BA. 1977a. Mushrooms in Yoruba mythology and medicinal practices, Eco. Bot. 31: 367-371

Oso BA. 1977b. Pleurotus tubes-regium from Nigeria. Mycologia, 67: 271-279.

Olawuyi OJ, Jonathan SG, Aina DA and Baysah GI. 2011. Studies on Anti-bacterial Activities of Fomes lignosus-A Nigerian Mushroom, Afr. J. Biomed. Res. 14: 57- 62.

Park YK, Lee HB, Jeon EJ, et al. 2004. Chaga mushroom extract inhibits oxidative DNA damage in human lymphocytes as assessed by comet assay. Biofactors. 21: 109-112.

Park YM, Won JH, Kim YH. et al. 2005. In vivo and in vitro anti-inflammatory and anti-nociceptive effects of the methanol extract of Inonotus obliquus. J. Ethnopharmacol. 101: 120-128.

Ren Z, Guo Z, Meydani SN and Wu D. 2008. White button mushroom enhances maturation of bone marrow-derived dendritic cells and their antigen presenting function in mice. The J. Nutrition, 138(3): 544-50.

Shi YL, James AE, Benzie IFF and Buswell JA. 2002. Mushroom-derived preparations in the prevention of H O -induced oxidative damage to cellular DNA. Teratogenesis, Carcinogenesis, Mutagenesis.

Sliva 1992. Pleurotus ostreatus inhibits proliferation of human breast and colon cancer cells through p53-dependent as well as p53-independent pathway. International Journal of Oncology. 1307-13. doi:10.3892/ijo_00000122.

Shin Y, Ramai Y and Terazawa M. 2001. Chemical constituents of Inonotus obliquus II: anew triterpene, 21, 24-cyclopentalanosta- 3b, 21,25-triol-8-ene from sclerotium. J Wood Sci. 47: 313-316.

Shi DH, Xu C, Guo BX. et al. 2008. Inhibition of soluble epoxide hydrolase by extracts derived from inflammation-treating Chinese medicinal herbs. Phytother Res. 22: 1264-1268.

Saar M. 1991. Fungi in Khanty folk medicine. JE thnopharmacol. 31: 175-179.

Sung HL, Hee SH and Jong WY. 2009. Antitumor Activity of Water Extract of a Mushroom, Inonotus obliquus, against HT-29 Human Colon Cancer Cells, Phytother. Res. DOI: 10.1002/ptr.

Song G and Du Q. 2010. Isolation of a polysaccharide with anticancer activity from Auricularia polytricha using high-speed countercurrent chromatography with an aqueous two-phase system. J Chromatogr A1217 (38):5930-4.

Taji S, Yamada T and In Y. 2007. Three new lanostane triterpenoids from Inonotus obliquus. Helv Chim Acta. 9: 2047-2057.

Takeujchi H, He P and Mooi LY. 2004. Reductive effect of hot-water extracts from wood year (Auricularia auricula-judae Quel.) on food intake and blood glucose concentration in genetically diabetic KK-Ay mice. J. Nutritional science and vitaminology. 50 (4): 300-4.

Vinogradov E, Petersen BO, Duus JAE and Wasser S. 2004 a. The structure of the glucuron oxylomannan produced by culinary-medicinal yellow brain mushroom, Tremella mesenterica Ritz.: Fr., Heterobasidiomycetes grown as one cell biomass in submerged culture. Carbohydrate Research. 339(8): 1483-9.

Vinogradov E, Petersen BO, Duus JO and Wasser SP. 2004b. The Isolation, Structure, and Applications of the Exocellular Heteropolysaccharide Glucuronoxylomannan Produced by Yellow Brain Mushroom Tremella mesenterica Ritz.: Fr. (Heterobasidiomycetes). International J. Medicinal Mushrooms. 6(4): 335.

Wasser SP and Weis AL. 1999a. Therapeutic effects of substances occurring in higher basidiomycetes mushrooms: a modern perspective, Crit. Rev. Immunol. 19: 65-96.

Wasser SP and Weis AL. 1999b. Medicinal properties of substances occurring in higher basidiomycete mushrooms: current perspectives (review), Int. J. Med. Mushrooms, 1: 31-62.

Wu X, Zheng S, Cui L. et al. 2010. Isolation and characterization of a novel ribonuclease from the pink oyster mushroom Pleurotus djamor. The J. General and Applied Microbiology. 56(3): 231-9.

Wasser SP. 2002. Medicinal mushrooms as a source of antitumor and immuno-modulating polysaccharides. Appl Microbiol Biotechnol 60: 258-274.

Wu D, Pae M, Ren Z. et al. 2007. Dietary supplementation with white button mushroom enhances natural killer cell activity in C57BL/6mice. The J. Nutrition, 137(6): 1472-7.

Yu L, Fernig DG, Smith JA et al. 1993. Reversible inhibition of proliferation of epithelial cell lines by Agaricus bisporus (edible mushroom) lectin. Cancer Research, 53(19): 4627-32.

Youn MJ, Kim JK and Park SY. 2008. Chaga mushroom (Inonotus obliquus) induces G0/G1arrest and apoptosis in human hepatoma HepG2 cells. World J Gastroenterol. 14: 511-517.

Zhang M, Cui SW, Cheung PCK and Wang Q. 2007. Antitumor polysaccharides from mushrooms: a review on their isolation process, structural characteristics and antitumor activity. Trends Food Sci. Tech. 18: 4-9.

Zusman I, Reifen R, Livni O, et al. 1997. Role of apoptosis, proliferating cell nuclear antigen and p53 protein in chemically induced colon cancer inrats fed corncob fiber treated with the fungus Pleurotus ostreatus. Anticancer research, 17(3C): 2105-13.

2.2.2 AYURVEDIC AND UNANI PLANTS [Sajib and Zakaria 2013, Zakaria et al 2014, Zakaria et al 2019]

2.2.2.1 INTRODUCTION

Ayurveda is a natural system of medicine that has been practiced in India for more than 5,000 years. Ayurveda is a Sanskit word made up of two components, ayush meaning life, and veda meaning knowledge or science. Hence, Ayurveda is the "science of life". It was developed by seers (rishis) through centuries of observation, experimentation, discussion, and meditation. The origin of ayurvedic medicine are recorded in the Atharva Veda, one of the four Vedic scriptures. For several thousand years, ayurvedic teachings were passed down orally from teacher to student. The first summary of these teachings was put into writing around 1500 B.C. The main sources of knowledge are the three Vedic classics Charaka Samhita, Susruta Samhita, and Ashtanga Hridaya (**Mishra et al 2008**). The main objectives of ayurveda are-

(i) to maintain and promote health by preventing physical, mental, and spiritual ailments; and

(ii) to cure disease through natural medicine, diet, and a regulated lifestyle.

In this system of medicine, more than 1200 plants, nearly 100 minerals and over 100 animal products are used. Due to considerable research on pharmacognosy, chemistry, pharmacology and clinical therapeutics, numerous drugs have entered into international pharmacopoeia. Ayurvedic medicines are of various types, to meet the diverse requirements in the treatment of illness. They are herbal teas, infusions, decoctions, tinctures, capsules and powders, infused oils, ointments, creams, lotions (**Kumar et al 2010**). In Bangladesh, there are many ayurvedic Medicine practice now-a-days like Draksharishta, Kanakasava and Jierbadyarishta. Draksharishta is made of grape juice and herbs, but minus the micro-filtration process that modern wineries use, and minus the stringent temperature monitoring in its making and storage. Even now Draksharishta isprescribed by ayurvedic physicians as medicine, 6 spoons of Draksharishta mixed in equal amount of water, with meals (**Chakranews 2011**). It improves appetite, relieves constipation, provides strength, induces good sleep and relieves fatigue. It is highly effectivein correcting the weakness of lungs, relieving respiratory infection (reduces inflammation of respiratory tract) and reduces cough in tuberculosis patients (Sandu 2012).On the other hand, Kanakasava is mainly used in the treatment of respiratory diseases but should only be taken under strict medical supervision, because it contains a toxic herb ingredient called Dattura.

Kanakasava contains 5–10 % of self-generated alcohol. The self-generated alcohol and the water present in the product acts as a media to deliver water and alcohol soluble active herbal components to the body. It is usually one of the common Ayurvedic medicine for asthma but also used in the treatment of all types of respiratory diseases, COPD, bronchitis, bleeding diseases, injuries and chronic fever, etc. It is a natural mucolytic and bronchodilator. It helps to relieve chest mucous congestion but sometimes it may result in side effects resembling datura poisoning (**Ayurmedinfo 2012**). Another ayurvedic medicine called, Jierbadyarishta, is a special formulation of several herbs and natural ingredients that helps to control acidity, flatulence and indigestion, that is it tones up the digestive system as a whole and also makes a fine balance between liver and gastric activities.

The Unani System of medicine pioneered in Greece and was developed by Arabs into an elaborate medical science based on the frame work of the teaching of Buqrat (Hippocrates) an Jalinoos (Galen). Since that time Unani medicine has been known as Greco-Arab medicine. This system is based on Hippocratic theory of four humours viz. blood, phlegm, yellow bile and black bile, and the four qualities of states of living human body like hot, cold, moist and dry. They are represented as earth, water, fire and air, the Greek ideas were put by Arabian physician as seven principles (Umoor-e-Tabbiya) and included, element (Arkan), temperament (Mizaj), humours (Akhlat), organs (Aaza), sprit (Arwah), faculties (Qowa) and functions (Afaal). In this system is it believed that, these principles are responsible for the body constitution and its health (**Husain et al 2010**). The World Health Organization (WHO) has recognized the Unani System of Medicine (USM) as an alternative system for the health care needs of human population. Alternative medicine is being practiced worldwide. Unani is one of the most well-known traditional medicine systems and draws on the ancient traditional systems of medicine of China, Egypt, India, Iraq, Persia and Syria. In fact Unani medicine and herbal products are gradually more being used in many countries where modern medicine is easily available. India has accepted it as one of the alternative health care system and has given it official status (**Husain et al 2010**). In Bangladesh, there are some unani medicine which includes- Alvasin and Carmina. Alvasin is a unique combination of valuable herbs for all types of cough and cold. Vasicine is the main ingredient of Bashok which has expectorant, bronchodilator, antitussive, mucolytic and antiallergic properties. Jastimadhu has antiallergic properties due to glycyrrhizic acid by inhibiting the secretion of histamin. Alvasin is effective and safe, well tolerated and non-sedative for both adults and children (Hamdard 2008). It is used for all types of cough (dry cough, smokers cough, hacking cough, whooping cough, tubercular cough, allergic cough), common cold, congestion of lungs, bronchitis, tonsillitis, bronchial asthma, hoarseness of voice, sore throat and influenza. On the other hand, Carmina helps in toning up the mucosa of the stomach and brings about a well-toleratedbalance between liver functions and gastric activities. Carmina is a special formulation of several natural ingredients that helps to control acidity, flatulence and indigestion (Hamdard 2008). The present study was designed to observe antimicrobial activities of the ayurvedic and unani medicines namely- Draksharishta, Kanakasava, Jierbadyarishta, Alvasin and Carmina. The current study was structured for screening of in-vitro antimicrobial activities of collected different samples by agar-well method, application of different concentration of these products against test organisms alone with the comparison against commercial antibiotics, determination of combined effect of antimicrobial activity of these testproducts and finally study the effect of antimicrobial activity of these test products against single and mixed microbial culture.

2.2.2.2 MATERIALS AND METHODS

Throughout the present study, reagent grade nutrient agar medium was used. The selected test medicines were collected from Ayurvedic laboratories Noakhali, Bangladesh. Both Gram-positive and Gram-negative strains of bacteria (*Bacillus subtilis, Enterococcus* sp., *Staphylococcus* sp., *Klebsiella*sp., *Salmonella typhi* , *E. coli, Acinetobacter* sp. and *Proteus* sp.) were used as the test organism to observe the antibacterial activity of the test product. These organisms were collected from the Microbiology research laboraroty, Department of Microbiology, Primeasia University, Dhaka, Bangladesh. Test antibiotics in this study are-Tetracycline (30µg/disc), Vancomycin (30µg/disc), Cefaclor (30µg/disc), Sulphamethoxazole / Trimethoprim (25µg/disc), Penicillin (10µg/disc), Clindamycin (2µg/disc), Rifampicin (5µg/disc), Amoxycillin (10µg/disc), Erythromycin (15 µg/disc) and Cloxacillin (5µg/disc). Agar wells of 5 mm in diameter, 4 mm deep and about 2 cm apart were punched in the NA agar with a sterile cork-borer. Approximately 40 µl of the test product were dropped into each well which filled them respectively to fullness. The setup was allowed to stabilize for 3h before being incubated at 37°C for 24h as described previously by Aibinu et al (**2007**).

2.2.2.2.1 Standardization of Inoculum and Assay of Sensitivity Pattern A standard stock of the bacteria isolates was prepared by suspending a loop full of each microbial growth in about 4 ml of distilled water. After incubation at 37°C for 12h, the turbidity was adjusted to be visually comparable with a 0.5 McFarland's standard giving a bacterial load of about 3.6×10^2cfu/mL. 1ml of each of two different microbial culture (from 3.6×10^2cfu/mL stock culture) were mixed together aseptically and used for assay the sensitivity pattern against test product.

2.2.2.2.2 Determination of MIC and MBC The MIC is the lowest concentration of the test sample or drug at which it shows the highest activity against microorganism. The minimum inhibitory concentration was determined by serial dilution techniques using nutrient broth media. The minimum inhibitory concentration of test product was determined against all of the selected microbes. Dilutions showing no visible growth for the MIC was sub-cultured and incubated at 37°C for 24h. The lowest concentration of test product yielding no growth on the NA plate was recorded as the MBC. Serial dilutions of test product 1:2, 1:4, 1:8 and 1:16 ml/ml were prepared and their antimicrobial activities were determined by agar well procedure.

2.2.2.2.3 Compare with Commercial Antibiotic of the Test Product Nutrient agar plates were prepared and incubated at 37°C for 24h to observed contamination. Then 2 Wells of 5mm in diameter, 4 mm deep and about 2cm apart were punched in the middle of the NA agar plate with a sterile cork-borer. Each Nutrient agar (NA) agar plate was uniformly seeded by means of sterile spreader dipped in the suspension and spread on the agar plate surface, and the plates left on the bench for excess fluid to be absorbed. Approximately 40μl of the test product and 1:2 concentration of the test product were dropped into each well which filled them respectively tofullness. Then ten selected commercial antibiotic disk were placed surround the well at equal distances on NA plate surface by Sterile forceps and incubated at 37°C for 24h. This procedure was applied for each test product and organism. The antimicrobial activities of the test product and commercial antibiotic were determined by measuring the diameter of the zones of inhibition in millimeter with transparent scale.

2.2.2.2.4 Combined Effect of Test Product against Selected Microbes Nutrient agar plates were prepared and incubated at 37°C for 24h to observed contamination. Then 3 Wells of 5mm in diameter, 4mm deep and about 2cm apart were punched at equal distances on NA agar plate with a sterile cork-borer. Each Nutrient agar (NA) agar plate was uniformly seeded by means of sterile spreader dipped in the suspension and spread on the agar plate surface, and the plates left on the bench for excess fluid to be absorbed. Each well was filled with combination of three of two test product such as one well was filled with combination of 20µl of Draksharishta and 20µl of Kanakasava, another well was filled with combination of 20µl of Draksharishta and 20µl of Jierbadyarishta, and rest one was filled with 20µl Kanakasava and 20µl of Jierbadyarishta. The plates were then incubated at 37°C for 24h. The antimicrobial activities of the combination of test productwere determined by measuring the diameter of the zones of inhibition in millimeter with transparent scale.

2.2.2.2.5 Determination of Effect of Test Product against Mixed Culture 1ml of stock solution (3.6×10^2cfu /ml) of each two test microbes were mixed in a sterile test tube. Three Wells of 5mm in diameter, 4mm deep and about 2cm apart were punched at equal distances on NA agar plate with a sterile cork-borer. Then mixture was spread by a spreader on the nutrient agar plate surface. Approximately 40µl of the test product were dropped into each well which filled them respectively to fullness. The plates were then incubated at 37°C for 24h. The antimicrobial activities of test product against mixed culture were determined by measuring the diameter of the zones of inhibition in millimeter with transparent scale.

2.2.2.3 RESULTS AND DISCUSSIONS

2.2.2.3.1 Antimicrobial Activity Ayurvedic Medicine namely- Draksharishta, Kanakasava, Jierbadyarishta were tested (agar well method) for their antimicrobial activety against all the selected microorganisms. In this antibacterial screening of test product were used at a concentration of 40µg/well. Unani Medicine namely- Alvasin and Carmina were tested (agar well method) for their antimicrobial activety against all the selected microorganisms. In this antibacterial screening of test product were used at a concentration of 40µg/well. All of test organisms showed sensitive against all test product (Draksharishta, Kanakasava, Jierbadyarishta). Jierbadyarishta showed 15mm highest zone of inhibition against*Proteus* sp.,*Enterococcus* sp. *Acinetobacter*sp., *Staphylococcus* sp., *Klebsiella* sp. where Draksharishta showed 15mm zone of inhibition against *Staphylococcus* sp., *Klebsiella*sp., *Salmonella typhi, Bacillus subtilis, Enterococcus* sp., *Acinetobacter* sp. and 17.5 and 10mm zone of inhibition against *Proteus* sp., *Escherichia coli* respectively. Kanakasava showed 18mm highest zone of inhibition against *Klebsiella* sp. (**Table 2.2.2.1**).

Table 2.2.2.1: Determination of Zone of Inhibition by Ayurvedic Medicine. [Sajib and Zakaria 2013, Zakaria et al 2014, Zakaria et al 2019]

Name of Organisms	Zone of Inhibition (mm)		
	Jierbadyarishta	Kanakasava	Draksharishta
Salmonella typhi	10	16	15
Proteus sp.	15	17.5	17.5
Acinetobacter sp.	15	15	15
E. coli	10	9	10
Staphylococcus sp.	15	13	15
Klebsiella sp.	15	18	15
Bacillus subtilis	10	12.5	15
Enterococcus sp.	15	15	15

Different concentration of Jierbadyarishta (Neat, 1:2, 1:4, 1:8 and 1:16) were applied in different selected microbes (3.6×10^2cfu/ml) and the zone of inhibition assayed. In **Table 2.2.2.2**, it was observed that upto 1:2 concentration showed better result and in the 1:4 the zone of inhibition started to decreased and 1:16 it showed almost negative result.

Table 2.2.2.2: Application of Jierbadyarishta in Different Concentration. [Sajib and Zakaria 2013, Zakaria et al 2014, Zakaria et al 2019].

Name of Organisms	Zone of Inhibition (mm)				
	Neat	1:2	1:4	1:8	1:16
Enterococcus sp.	15	9	5	5	0
Salmonella typhi	10	7.5	0	0	0
Bacillus subtilis	10	6	0	0	0
Acinetobacter sp.	15	0	0	0	0
E. coli	10	0	0	0	0
Klebsiella sp.	15	5	0	0	0
Staphylococcus sp.	15	4.5	0	0	0
Proteus sp.	15	5	0	0	0

In case of Kanakasava, it was observed that all the different concentrations alone with Neat showed satisfactory results whereas in case of Draksharishta, satisfactory zone of inhibition were observed only in concentration at neat and 1:2, in the 1:4 the zone of inhibition started to decreased and 1:16 it showed almost negative result (**Table 2.2.2.3 &2.2.2.4**).

Table 2.2.2.3: Application of Kanakasava in Different Concentration. [Sajib and Zakaria 2013]

Name of Organisms	Zone of Inhibition (mm)				
	Neat	1:2	1:4	1:8	1:16
Enterococcus sp.	15	8	6	0	0
Salmonella typhi	16	5	0	0	0
Bacillus subtilis	12.5	7.5	0	0	0
Acinetobacter sp.	17.5	8	0	0	0
E. coli	9	7.5	0	0	0
Klebsiella sp.	18	0	0	0	0
Staphylococcus sp.	13	4	0	0	0
Proteus sp.	17.5	0	0	0	0

Table 2.2.2.4: Application of Draksharishta in Different Concentration. [Sajib and Zakaria 2013].

Name of Organisms	Zone of Inhibition (mm)				
	Neat	1:2	1:4	1:8	1:16
Enterococcus sp.	15	10	0	0	0
Salmonella typhi	15	7	5	0	0
Bacillus subtilis	15	6	0	0	0
Acinetobacter sp.	15	7.5	0	0	0
E. coli	10	5	0	0	0
Klebsiella sp.	15	0	0	0	0
Staphylococcus sp.	15	5	0	0	0
Proteus sp.	17.5	4	0	0	0

Table 2.2.2.5: Effect of Test Products against Mixed Cultures. [Sajib and Zakaria 2013].

Name of Organisms	Zone of Inhibition (mm)		
	Jierbadyarishta	Kanakasava	Draksharishta
Klebsiella sp. + *Salmonella* sp.	10	12	9
E. coli + *Salmonella typhi*	9	10	12
E. coli +*Enterococcus* sp.	11	12	10
B. cereus + *Staphylococcus* sp.	9	11	10
B. cereus + *Salmonella* sp.	9	8	10
Enterococcus sp. +*B. cereus*	8	12	11
Proteus sp. +*Acinetobacter* sp.	10	10	9

Table 2.2.2.5showed the antimicrobial activity of Alvasin were 8, 6, 6. 5, 5, 5, 8, 5, and 7mm zone of inhibition and the antimicrobial activity of Carmina were 7.5, 9, 8, 0, 9, 7.5, 5, and 13mm zone of inhibition were observed against *Salmonella typhi, Proteus* sp., *Acinetobacte*r sp.,*E.coli, Staphylococcus* sp., *Klebsiella* sp.,*Bacillus subtilis* and*Enterococcus* sp.,respectively.The results for the antibacterial screening have shown that all tested Ayurvedic Medicines in this present study have antibacterial activity. The results of the inhibition of bacterial growth have shown that tested Ayurvedic Medicines are active at high concentration and inactive at very low concentrations. Thus the study may suggest that the inhibition of bacterial growth activity of the products is dose dependent. It was observed that all the tested Ayurvedic medicines showed

remarkable activity against all the tested organisms with the highest activity found on *Klebsiella*sp. (18mm zone diameter) and lowest activity found on *E.coli* (9mm zone diameter) against Kanakasava at 40µl/well concentration. Jierbadyarishta was also showed almost same antimicrobial activity (15mm) against all the tested organisms. In case of Draksharishta, it showed good antimicrobial activity against all the selected microbes with the highest activity found on*Proteus* sp. (17.5mm zone diameter) and lowest activity against *E.coli* (10mm zone diameter) at 40µl/well concentration. On the other hand, Tiwari et al (2011) worked on the same (Draksharishta) and found significant reduction in atherogenic index as compared to high fat diet fed control group which strongly supports antiatherosclerotic property of Draksharishta. Tiwari et al (2011) also showed that Draksharishta and glibenclamide exhibited significant antihyperglycemic activity in alloxan induced diabetic rats without causing significant change in body weight. In this present study it has shown that among all selected microbes, *Proteus* sp. showed highest sensitivity against all the tested Ayurvedic medicines where*E. coli* showed lowest sensitivity. In case of combination of Kanakasava and Jierbadyarishta range of zone of inhibition against the selected microbes was observed from 8-12mm, where 7-10mm zone of inhibition was observed in case of combination of Draksharishta and Jierbadyarishta and 6-8mm zone of inhibition was observed in case of combination of Kanakasava and Draksharishta. On the other hand, in this present study it has shown that Unani medicines (Alvasin and Carmina syrup) have less antimicrobial activity than local Ayurvedic Medicines against all the selected organisms. It was observed that Carmina syrup showed highest antimicrobial activity (13mm zone diameter) against *Enterococcus* sp. and no activity against *E.coli* at 40µl/well concentration. In case of Alvasin syrup, range of zone of inhibition against the selected microbes were observed from 5-8mm.

2.2.2.3.2 Determination of MIC and MBC In **Figure2.2.2.1**, it was observed that the MIC of Jierbadyarishta against *Salmonella typhi, Proteus* sp., *Acinetobacter* sp. and *E. coli* were 1µl/ml whereas MIC against *Klebsiella* sp., *Bacillus subtilis, Staphylococcus* sp. and *Enterococcus* sp. were 0.1µl/ml. The MBC of Jierbadyarishta against*Salmonella typhi* and *Klebsiella* sp. were 10µl/ml and 0.1µl/ml whereas 1µl/ml MBC was observed against all other organisms. In case of Kanakasava, MIC against *Staphylococcus* sp. were 0.1µl/ml and MIC of 1µl/ml were observed in case of other microbes (**Fig2.2.2.2**). On the other hand, MBC of 1µl/ml were observed against *Salmonella typhi* and *Enterococcus* sp., 10µl/ml was observed against *E. coli, Proteus* sp., *Acinetobacter*sp. and 0.1µl/ml was observed against all other microbes.

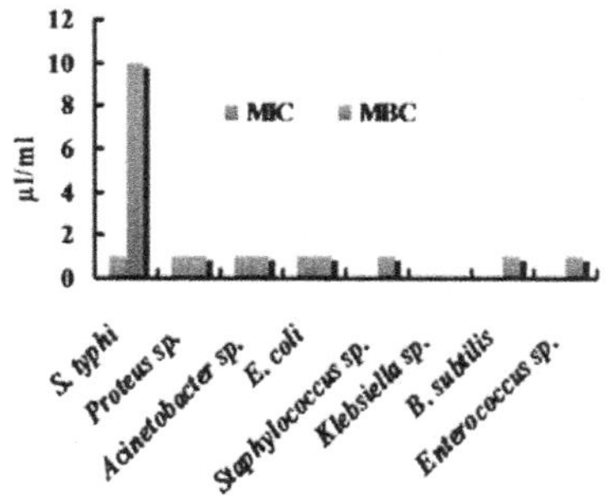

Figure 2.2.2.1: MIC and MBC of Jierbadyarishta. [Sajib and Zakaria 2013]

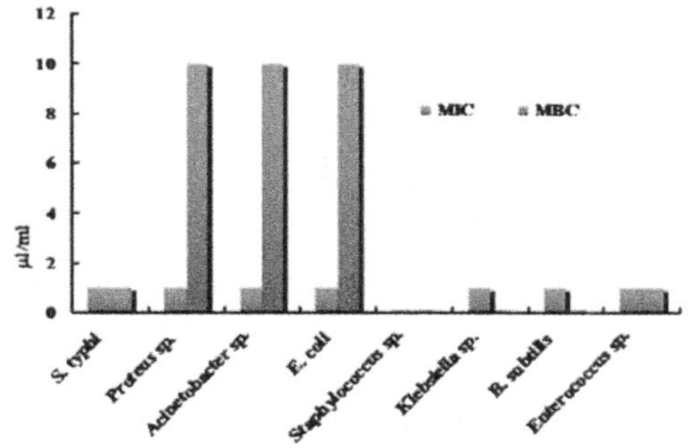

Figure 2.2.2.2: MIC and MBC of Kanakasava. [Sajib and Zakaria 2013]

In case of Draksharishta, MIC against *Staphylococcus* sp,, *Klebsiella* sp., *Salmonella typhi, Proteus* sp., *Enterococcus* sp. were 0.1µl/ml and MIC of 1µl/ml were observed in case of other microbes (**Fig2.2.2.3**). On the other hand, MBC of 1µl/ml were observed against *Klebsiella* sp., *Proteus* sp., and *Acinetobacter* sp.; 0.1µl/ml was observed against *Enterococcus* sp., and 10µl/ml was observed against all other microbes. In present study, it was observed that the MIC of Jierbadyarishta against *Salmonella typhi, Proteus* sp., *Acinetobacter* sp. and *E. coli* were 1µl/ml whereas MIC against *Klebsiella* sp., *Bacillus subtilis, Staphylococcus* sp., and *Enterococcus* sp. were 0.1µl/ml. The MBC of Jierbadyarishta against *Salmonella typhi* and *Klebsiella* sp. were 10µl/ml and 0.1µl/ml whereas 1µl/ml MBC was observed against all other organisms. In case of Kanakasava,

MIC against *Staphylococcus* sp. were 0.1µl/ml and MIC of 1µl/ml were observed in case of other microbes. On the other hand, MBC of 1µl/ml were observed against*Salmonella typhi*and*Enterococcus* sp.,10µl/ml was observed against*E. coli*,*Proteus* sp., *Acinetobacter* sp. and 0.1µl/ml was observed against all other microbes. In case of Draksharishta, MIC against*Staphylococcus* sp,, *Klebsiella* sp., *Salmonella typhi, Proteus* sp., *Enterococcus* sp. were 0.1µl/ml and MIC of 1µl/ml were observed in case of other microbes. On the other hand, MBC of 1µl/ml were observed against *Klebsiella* sp., *Proteus* sp., and*Acinetobacter* sp.; 0.1µl/ml was observed against*Enterococcus* sp., and 10µl/ml was observed against all other microbes. Tiwari et al (2011) observed that Draksharishta inhibited the ferrous sulphate induced lipid peroxidation in a dose dependent manner and showed inhibitory concentration (IC_{50}) value 230.03µg/ml.

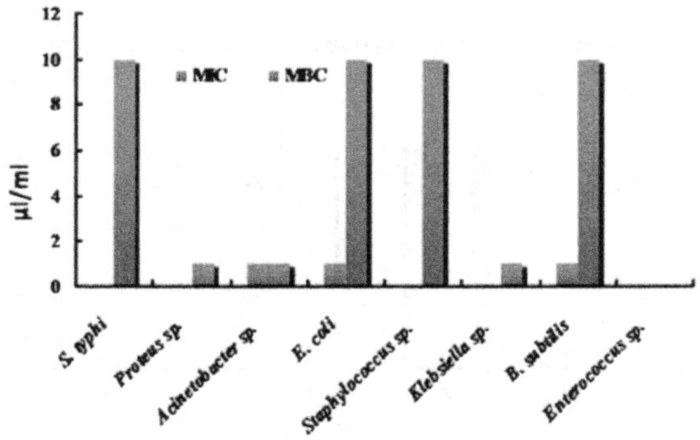

Figure 2.2.2.3: MIC and MBC of Draksharishta. [Sajib and Zakaria 2013]

2.2.2.3.3Combined Effect of Test Product against Selected Microbes Combination effect of different test medicines (Kanakasava+Jierbadyarishta, Draksharishta+Jierbadyarishta, and Kanakasava+Draksharishta) showed satisfactory results against all the selected microbes. The zone of inhibition of Kanakasava & Jierbadyarishta were found 8, 11, 12, 11, 9, 10, 10, 8 (**Fig2.2.2.4A**) whereas zone of inhibition of Draksharishtaand Jierbadyarishta were 10, 10, 9, 9, 7, 10, 9, 10 (**Fig2.2.2.4B**) and zone of inhibition of Kanakasava and Draksharishta were 7, 6, 8, 7, 7, 6, 8, 6 (**Fig 2.2.2.5**) showed against*E.coli*,*S. typhi, Proteus* sp., *Acinetobacter* sp., *Staphylococcus* sp., *Klebsiella* sp., *B.subtilis* and *Enterococcus* sp., respectively.

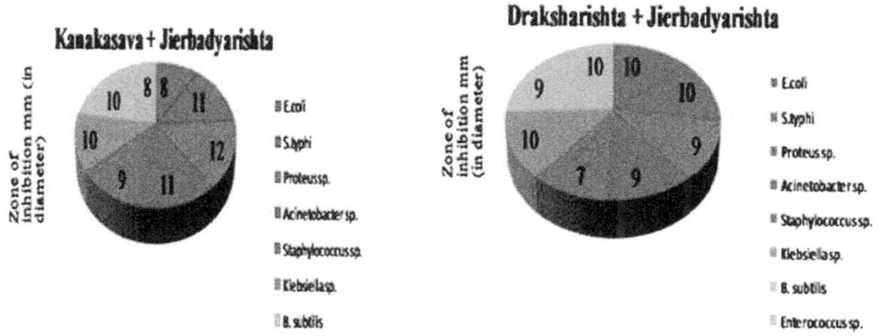

Figure 2.2.2.4: Combined Effect of (A) Kanakasava & Jierbadyarishta (B) Draksharishta + Jierbadyarishta. [Sajib and Zakaria 2013]

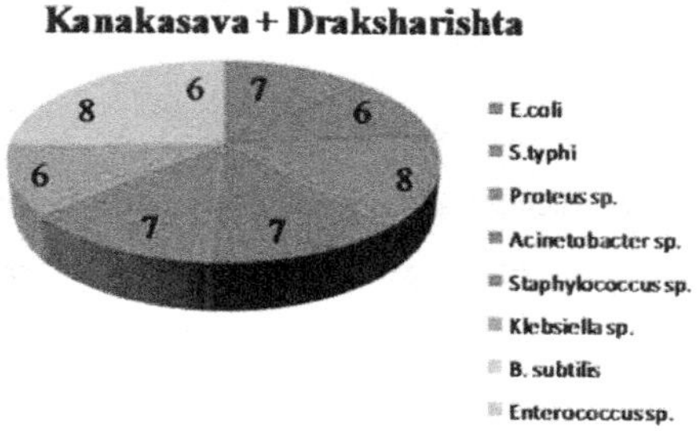

Figure 2.2.2.5: Combined Effect of Kanakasava & Draksharishta. [Sajib and Zakaria 2013]

2.2.2.3.4Determination of Effect of Test Products against Mixed Culture In **Table 2.2.2.6**, Jierbadyarishta was effective against all mixed microbial cultures and the zone of inhibition against *Klebsiella* sp.+*Salmonella* sp., *E. coli+Salmonella* typhi, *E. coli+Enterococcus* sp., *B. cereus+Staphylococcus* sp., *B. cereus+Salmonella* sp., *Enterococcus* sp.+*B. cereus, Proteus*sp.+*Acinetobacter* sp. were found 10, 9, 11, 9, 9, 8, and 10 respectively. In case of Kanakasava, satisfactory results were found against*Klebsiella*sp.+*Salmonella* sp., *E. coli+Salmonella typhi, E. coli+Enterococcus* sp., *B. cereus+Staphylococcus* spp., *B. cereus+Salmonella* sp., *Enterococcus*sp.+*B. cereus, Proteus+Acinetobacter* sp. where the zone of inhibition were 12, 10, 12, 11, 8, 12 and 10, respectively (**Table 2.2.2.7**). In case of Draksharishta, satisfactory results were found against *Klebsiella*sp.+*Salmonella* sp., *E. coli+Salmonella typhi, E. coli+Enterococcus* sp., *B. cereus+Staphylococcus* sp., *B. cereus + Salmonella* sp., *Enterococcus* sp.+*B. cereus, Proteus* sp.+*Acinetobacter* sp. where the zone of inhibition were 9, 12, 10, 10, 10, 11 and 9, respectively (**Table 2.2.2.1**). **Table 2.2.2.6** shows that all the test products (Draksharishta,

Kanakasava, Jierbadyarishta) showed satisfactory results aginst all mixed microbial cultures (*Klebsiella* sp.+*Salmonella*sp., *E. coli*+*Salmonella typhi*, *E. coli*+*Enterococcus* sp., *B. cereus*+*Staphylococcus* sp., *B. cereus* + *Salmonella* sp., *Enterococcus* sp.+*B. cereus*, *Proteus* sp.+*Acinetobacter*sp.).

Table 2.2.2.6: Antimicrobial Activity of Unani Medicine. [Sajib and Zakaria 2013, Zakaria et al 2014, Zakaria et al 2019]

Name of Organisms	Zone of Inhibition (mm)	
	Alvasin	Carmina
Salmonella typhi	8	7.5
Proteus sp.	6	9
Acinetobacter sp.	6.5	8
E.coli	5	0
Staphylococcus sp.	5	9
Klebsiella sp.	8	7.5
Bacillus subtilis	5	5
Enterococcus sp.	7	13

2.2.2.3.5Comparison of Test Medicine Products with Commercial Antibiotics In case of *E. coli* the test products namely- Jierbadyarishta, Kanakasava, Draksharishta showed 10, 9, and 10mm zone of inhibition respectively whereas 14, 8mm zone of inhibition were observed against Sulphamethoxazole/Trimethoprim, and Erythromycin and respectively. At the time of *S. typhi* 10, 16, 15mm zone of inhibition were showed against Jierbadyarishta, Kanakasava, Draksharishta,respectively whereas 12, 8, and 11mm zone of inhibition were observed against Sulphamethoxazole/Trimethoprim, Erythromycin and Cloxacillin respectively. In case of *Proteus* sp. highest zone of inhibition was observed 15, 17.5 and 17.5mm against Jierbadyarishta, Kanakasava, Draksharishta whereas 13, 9, 8, and 9.5 clear zone were observed against Sulphamethoxazole/Trimethoprim, Penicillin, Amoxycillin and Cloxacillin respectively. At the time of *Acinetobacter* sp. clear zone of inhibition of 15, 17.5 and 15mm were observed against Jierbadyarishta, Kanakasava, Draksharishta respectively whereas zone of inhibition by Sulphamethoxazole/Trimethoprim, Rifampicin, and Erythromycin were 14, 6 and 7mm respectively against *Acinetobacter* sp. In case of *Staphylococcus* sp. the test products namely- Jierbadyarishta, Kanakasava, Draksharishta showed 15, 13, and 15mm zone of inhibition, respectively; whereas 17, 6.5, 9 and 8mm zone of inhibition were observed against Clindamycin, Rifampicin, Amoxicillin and Tetracycline respectively. At the time of *Klebsiella* sp. 15, 18 and 15mm zone of inhibition were showed against Jierbadyarishta, Kanakasava, Draksharishtarespectively whereas 12, 7.5, and 9mm zone of inhibition were observed against Penicillin, Clindamycin and Rifampicin respectively. In case of *B. subtilis* highest zone of inhibition was observed 10, 12.5, and 15mm against Jierbadyarishta, Kanakasava, Draksharishta whereas 9, 9, 7.5, 14, 7.5 and 12 mm clear zone were observed against Sulphamethoxazole/Trimethoprim, Penicillin, Clindamycin, Rifampicin, Amoxycillin and Erythromycin respectively. At the time of *Enterococcus* sp. clear zone of inhibition of 15mm were observed against Jierbadyarishta, Kanakasava, Draksharishta whereas zone of inhibition by Clindamycin, Rifampicin, Amoxycillin, Tetracycline, Cefaclor and Vancomycin were 14, 9, 7.5, 9, 7.5 and 12mm, respectively against *Enterococcus* sp. (**Fig2.2.2.6** and **2.2.2.7**).

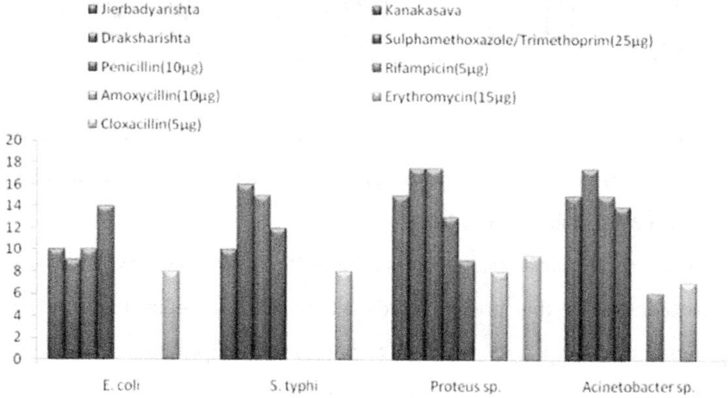

Figure 2.2.2.6: Comparison of Zone of Inhibition of Test Products and CommercialAntibiotics against *E. Coli, S. Typi, Proteus* sp. and *Acinetobacter* sp.[Sajib and Zakaria 2013, Zakaria et al 2014, Zakaria et al 2019]

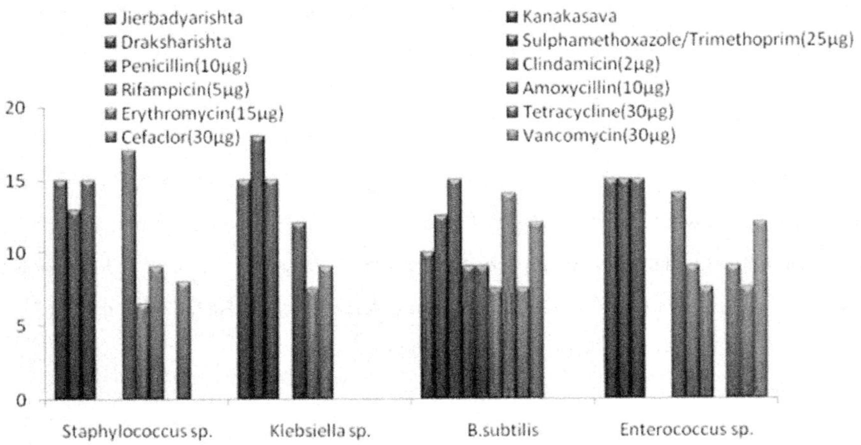

Figure 2.2.2.7: Comparison of Zone of Inhibition of Test Products and Commercial Antibiotics against *Staphylococcus* sp., *Klebsiella* sp., *B. Subtilis* and *Enterococcus* sp.[Sajib and Zakaria 2013, Zakaria et al 2014, Zakaria et al 2019]

Satisfactory results were observed in all the tested ayurvedic medicines as compared with commercial antibiotics. In this present study all of the ayurvedic medicines (Draksharishta, Kanakasava, Jierbadyarishta) tasted showed more effective antimicrobial agent than unani medicine (Alvasin and Carmina syrup) against all the tested organisms in controlling their growth in vitro in culture condition. They all have a bacteristatic and bactericidal activity when tested in vitro using selected neat sample. In this project it was interestingly found that all of the tested ayurvedic medicines have almost twice antimicrobial activity than unani medicine, even whereas most commercial antibiotic has a little response against all the tasted organisms. The microbes that were selected in this present study can cause different types of diseases. As if the tested ayurvedic medicines were more effective than commercial antibiotics like Sulphamethoxazole/Trimethoprim, Erythromycin, Tetracycline, Penicillin and they can be successfully used for the treatment of various diseases. The tested local ayurvedic medicines can also be preferred inested of the area of unani medicine application. More importantly there is need for detailed scientific study of traditional practices to ensure the valuable therapeutic knowledge of tested ayurvedic medicine and also to provide scientific evidencefor their efficacies. Since the tested ayurvedic medicines are made from natural plant products which are available in our country sides and having traditional medical use, development of modern drugs can be contributed with ayurvedic composition to cure very common diseases in our country at cheap rate especially for the people of rural areas. In fine the time has come to get turn into ayurvedic medicine application which has no side effects. For the last few years the mass awareness tends to ayurvedic medicine usage and increasing day by day.

2.2.2.4 REFERENCES

Ayurmedinfo (ayurmedinfo.com/.../kanakasava-dose-ingredients-side-effects-benefits)

Ayurblog(www.ayurblog.com/2012/01/draksharishta-ayurvedic-winefor.html).

Aibinu I, Adenipekum T, Adelowotan T, Ogunsanya T and Odugbemi T. 2007.Evaluation of the antimicrobial properties of different parts of Citrus aurantifolia (lime fruit) as used locally. Afr. J. Trad. Complem. Alter. Med. 4(2): 185-195.

Chakranews(http://www.chakranews.com/draksharishta-grape-wine-and-other-ayurvedic-wines-used-originally-as-medicine/1660).

Husain A, Sofi GD, Tajuddin DR and Kumar N. 2010. Unani system of medicine introduction and challenges. Medical Journal of Islamic World Academy of Sciences, 18(1): 27-30.

Hamdard (www.hamdard.com.bd/products/products-list/87-alvasin).

Hamdard (www.hamdard.com.bd/products/products-list/87-carmina).

Mishra K, Ravi B and Vinjamury S. 2008. Ayurvedic Medicine. Caring Ambassadors Program, Inc. 153-165.

National center for complementery and alternative medicine. 2005. 1-12. (Web site: nccam.nih.gov).

Praveen KSV, Sandeep M, Kamal D, Nishanth BC, Megharaj HK, Prashith Kekuda TR and Gurucharan DN. 2010. Antibacterial and Anthelmintic activity of selected fermented Ayurvedic herbal formulations. 2(7): 347-348.

Tiwari P, Patel RK. 2011. Evaulation of antihyperlipidemic activity of Draksharishta prepared by traditional and modern methods in hyperlipidemic rats. Pharmacologyonline 2: 576-585.

Sandu (www.sandu.in/draksharishta.html).

Wikipedia (en. wikipedia.org/wiki/unani).

2.2.3 TEA PLANTS[Zakaria 2013, Zakaria et al 2014, Zakaria et al 2019, Md. W. Siddiqui et al 2016]

2.2.3.1 Introduction

Plant materials are widely used in developed and developing countries as home remedies, over the counter drug products and raw materials of the pharmaceutical industry. They represent a substantial proportion of the global drug market (**Kunle et al 2012**). Many complementary and alternative medicines have increased their popularity in recent decades (**Kyungseop 2017**). According to previous studies(**Martins 2013, Napatt et al 2019**), tea (*Camellia sinensis*) isthe second most consumed beverage in the world next to water. Reygaert (**2017**) reported that tea belongs to the familyTheaceae, and is a product made by processing the leaves or buds. All tea varieties such as, green, oolong and black tea, are harvested from this species. Tea extracts work as effective treatments for patients who suffer from damaged skin following radiation treatment for cancer. Several previous studies (**Priyanka et al2012, Ohishi et al 2016, Arina et al2017**)revealed that this mightpartly be due to the anti-inflammatory propertiesof tea. Moreover, Arina et al (**2017**) added that tea acts at the cellular level by inhibiting the inflammatory pathways, and lowering the release of pro-inflammatory cytokines including; IL-1beta, IL-6, IL-8, TNF alpha, PGE2,and white blood cells in human cell lines. Green tea extract had been shown to be potent in neuro-protectant (Orly et al 2009). A study conducted by Qin et al (**2012**) suggested that the intracellular Amyloid beta (iAβ)-induced toxicity was significantly inhibited by green tea polyphenols in cultured rat primary cortical neurons. Previous studies (**Li et al 2007, Qin et al 2012**) reported that the intracellular iAβ toxicity is related to the development of Alzheimer's disease. In another study of Cong et al(**2016**), it was explored that green tea polyphenols inhibit glutamate-induced neurotoxicity which is linked to neurodegenerative diseases.

2.2.3.2 Chemical composition of tea*C. sinensis* mainly comprises phenolic compounds (cate-chins and flavonoids), specially alkaloids, such as caffeine and theophylline, volatile oils, polysaccharides, amino acids, lipids, vitamins, inorganic elements, for instance aluminium, fluorine, and manganese, among others. Caffeine is the main alkaloid present in tea that accounts for approximately2% of the plant composition. **Figure 2.2.3.1** illustrates some of the phenolic compounds found in tea extracts.Tea contains a group of compounds, particularly catechins,known for their antioxidant-related effects. The major tea catechins are (¡)-epicatechin

(EC), (¡)-epigallocatechin, and (¡)-epigallo-catechin gallate. Among these, EGCG is the most luxuriantcomponent in tea extract and the most potent chemical testedfor biological activity. These phenolics may account for asmuch as 30% of the dry weight of fresh tea leaves (FTL). Many compounds of tea have beenidentified, such as flown-3-ols, flavonols, gallic acid, theaflavins (TF), purine alkaloids, and amino acids. The main flavan-3-ols in tea are (¡)-epicatechin and its gallate derivatives, which will be partlyreplaced by theaflavins (theaflavin-3-gallate, theaflavin-30-gallate, and theaflavin-3-30-digallate) and thearubigins (TR) afterenzymatic oxidation (called tea fermentation). Major flavonols are quercetin, kaempferol, and myricetin, which not only affect the astringent tasteof tea but also its color. Flavonols are usually found in tea bound to sugars as O-glycosides, such as quercetin-3-O-glucoside, quercetin-3-O-galactoside, quercetin-3-O-rutinoside, and kaempferol-3-O-rutinoside (**Aherne and O'Brien, 2002**). A standardized profiling method based on high-performance liquid chromatographycombined with ultraviolet (UV) and mass spectrometric (MS)detection was established by Wu et al (**2012**) to analyze thephenolic compounds of selected tea varieties used formanufacturing green, black, and Oolong teas. The compositionand content include flavonol, flavones glycosides, phenolicacids, and purine alkaloids. Each tea variety had a unique chemical profile. The authors concluded that tea varieties with a highcontent of total catechins seem to be more suitable for fermented tea such as Oolong or black tea.Although both black and green teas contain similar amounts offlavonoids, their chemical structures are different. Green teacontains more catechins (simple flavonoids). Leaf oxidationduring black tea processing converts these simple flavonoidsinto aflavins and arubigins.

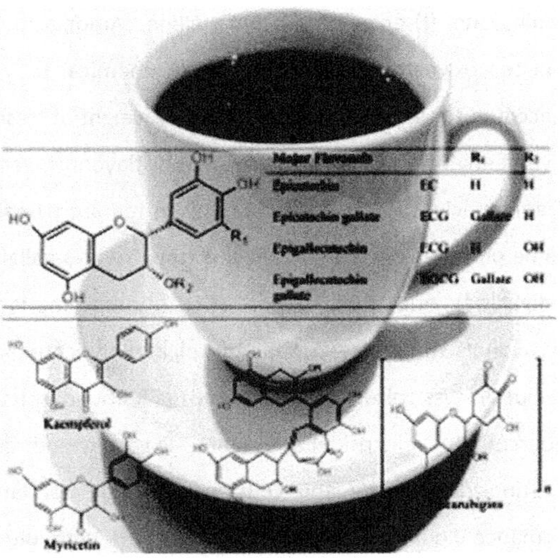

Figure 2.2.3.1 Polyphenolic compounds found in green and black tea extracts. [Md. W. Siddiqui et al 2016]

2.2.3.3 The effects and mechanism of action of tea Tea contains several naturally occurring dietary polyphenols such as catechins, which possess anti-carcinogenic activity that act as effective chemo-preventive agents against the initiation, promotion and progression stages of multistage carcinogenesis. Usually, all the four types of tea could lower the total cholesterol levels (**Aleksandra et al 2016**). Black tea has only about one-sixth of the antioxidant power of greentea owing mainly to the former's low levels of catechins (**Anton and Sheila 2003**). On the other hand, Duh et al (**2004**) reported that although the polyphenolic compositions of all the four types of tea were dramatically different, they have nearly identical levels of antioxidant potential. Agnieszka andWilfried (**2014**) suggested that the strong antioxidant catechins in green tea are destroyed by the fermentationreactions, leading toolong, black and Puerteas. Previously Jie et al (**2006**) revealed that there is a compensatory gain the antioxidant activity of tea due to the new compounds produced. Several years before, Chen et al (**2005**) discovered that extracts of Puer and black teas were effective inhibitors of the virus protein-destroying enzyme (called 3CLPro), responsible for Severe Acute Respiratory Syndrome (SARS) In addition, they identified three compounds found in fully fermented teas mainly; tannic acid, theaflavin-3'-gallate and the aflavin-3,3'-digallate, acting as effective 3CLPro inhibitors. In another study, Kuo et al (**2005**) reported that Puer and oolong significantly lowered the triglyceride levels, whereas green and black did not. Previously, (**Zhao et al 2002, Blanco et al 2005**) reported that the properties of green tea inhibit the bacterial growth, which are mainly related to their polyphenolic components including; epicatechin, epicatechin gallate, epigallocatechin, and epigallocatechin gallate, against various Gram-positive and Gram-negative bacteria. Moreover, Zhao et al (**2002**); Hu et al (**2002**); Stapleton et al (**2004**); Cho et al(**2008**) added that green tea has a synergistic effect with β-lactam antibiotics against Methicillin-resistant*Staphylococcus aureus* (MRSA). In a previous study, Zhao et al (**2001**) revealed that the main components of tea as polyphenols, epigallocatechin gallate, can reverse the methicillin resistance of MRSA by inhibiting the synthesis of Penicillin-binding Protein-2 (PBP2), and mayalso prevent adhesion to the mammalian epithelial cells (HEp-2) without any probable alteration (**Janecki and Kolodziej 2010, Sharma et al 2012**). It has been proposed by Lee et al (**2009**) that green tea leaves extract can prevent the attachment of pathogenic bacteria on the host cell membrane, which in turn acts as a potential anti-adhesive agent. Chung et al (**2003**) previously documented that green tea extract may affect the activity of dihydrofolate reductase, an enzyme that is needed by the pathogenic bacteria to synthesize purine and pyrimidine, as well as to increase the thickness of the epidermis.

Antimicrobial mode of action of plant extract is attributableto their phenolic compounds present. As a defensive functionin response to microbial infection, plants synthesize phenoliccompounds. It is therefore possiblethat they can act as effective antimicrobial substances againsta wide array of microorganisms. The biologically active molecules of plant extract are pondered as antimicrobial agents.However, the antimicrobial activity of plant extracts dependson not only phenolic compounds but also the property isdonated by the presence of different secondary metabolitessuch as hydroxyl groups of active constituents. Phenols as catechin acton different bacterial strains belonging to different species(*Escherichia coli, Bordetella bronchiseptica, Serratia marcescens, Klebsiella pneumonie, Salmonella choleraesis, Pseudomonas aeruginosa, Staphilococcus aureus*, and *Bacillussubtilis*) by generating hydrogen peroxide and by altering thepermeability of microbial membrane.This may also be attributable to the ability of these substancesto bind to bacterial adhesions, and by doing so they disturb theavailability of receptors on the surface.Some reports indicated that active constituents might attackcell wall and cell membrane, thereby destroying their permeability barrier and causing the release of intracellular constituents such as ribose and sodium glutamate. In addition, theyinterfere with electron transport, nutrient uptake, protein andnucleic acid synthesis, and enzyme activity, leading to theinhibition of bacterial growth (**Siddiqui et al 2016**).The inhibitory effects of tea, as other plant extracts, againstmicroorganism might be related to the role of mineral contents, active and effective phytochemical constituents. Thepresence of micronutrients and biologically active constituentsin plant extract usually interfere with growth and metabolismas well as destruction of microorganisms. Phenolic compounds also enhance antimicrobial activity by increasing the retention of certain minerals such as copper, magnesium, zinc, and Fe. The antimicrobial effect of teais based on the denaturation of membrane proteins, resulting inouter membrane disruption, subsequent KCleakage, respirationinhibition, and cell lysis. The phenolic extracts are also able toinhibit the synthesis of DNA, RNA, proteins, and polysaccharides in fungal and bacterial cells. Microorganisms stressed byexposure to phenols up-regulate proteins related to defensivemechanisms, which look after cells as simultaneously down-regulating various metabolic and biosynthetic proteinsinvolved, for example, in amino acid and protein synthesis aswell as phospholipid, carbon, and energy metabolism. Furthermore, phenolics interferewith bacterial quorum sensing, i.e., the fabrication of small signal molecules by bacterial cells of *Escherichia coli, Pseudomonas putida* and *Burkholderia cepacia*, which set off theexponential growth of bacterial population (**Siddiqui et al 2016**).

2.2.3.4 The tea tree oil (TTO) Recently, TTO has gained a good reputation as a safe, natural and effective antiseptic, andis employed significantly due to its antimicrobial properties. During the study of Carson et al (**2006**), it was reported that TTO is incorporated as the active ingredient in many topical formulations used to treat cutaneous infections. The chemical composition of TTO has been well defined, and it consists largely of cyclic monoterpenes of which about 50% areoxygenated, whereas the other 50% are hydrocarbons (**Cox et al 2000, Carson et al 2006**). TTO exhibits a broad-spectrum of antimicrobial activity which can be primarily attributed to the presence of terpinen-4-ol. Previously, Cox et al (**2000**) observed that the essential oil of *M. alternifolia* exhibited broad-spectrum antimicrobial activity, while TTO caused leakage of K ion in *E. coli, Staphylococcus aureus* and*Candida albicans*. Moreover, it has the ability to disrupt the permeability barrier of the cell membrane structures leading to the loss of chemiosmotic control, which is the most likely cause of its lethal action at minimum inhibitory levels. Hammer et al(**2000**) stated thatgerm tube formation (GTF) by*C. albicans* is affected by the presence of and/ or the pre-exposure to sub-inhibitory concentrations of TTO. In addition, TTO has been suggested by Hammer et al (**2000, 2003**) as a potential agent for MRSA decolonization. In a previous study, Hammer et al (1996) suggested that TTO could be useful for removing the transient skin flora, while suppressing but still maintaining the resident flora that acts as a natural defense against colonization by other pathogenic bacteria. In spite of having similar minimum inhibitory concentrations (MIC)/ minimum bactericidal concentrations (MBC) values, the microorganisms studied by Cox et al (**2000**) expressed obvious differences in their susceptibility to TTO. Previous work of Messager et al (**2005**) showed that the hand-cleansing formulations contain 5% TTO and 10% alcohol, or a solution of 5% TTO in water, which are more effective than soft soap in removing bacteria from the surface of the hand. Aaron et al (**2006**) studied the in vitro antibacterial activity of TTO against clinical skin isolates of methicillin-resistant *S. aureus* (MRSA), methicillin sensitive *S. aureus* (MSSA), and coagulase-negative Staphylococci (CoNS) growing as biofilms, and concluded that TTO exerted a greater bactericidal potential against the biofilm-growing MRSA and MSSA isolates, than against some of the biofilm-growing CoNS isolates.

In search of potent plant extracts, food grade additives thatcan effectively inhibit food borne pathogens having antioxidantproperties is a continuing challenge and an opportunity. Researchers are increasinglyrecognizing plant remedies as a very important low-cost alternative to industrially produced antibiotics that are not available to all who need them because of their high price. Microbes such as *Staphylococcus aureus, Salmonella typhi*, and *Pseudomonas aeruginosa* are the frequent pathogens of human infection. *S. aureus* is an opportunistic pathogen of human skin. *S. typhi* is an enteric pathogen involved in typhoid and enteric fever pathogenicity. *P. aeruginosa* is a pathogen associated with pyogenic infection and urinary tract infection. These microorganisms are highly pathogenic and the rate of prevalence of infection caused by these microorganisms has considerably risen in recent years. The antimicrobial activity of tea and tea extract is greatly affected by the nature of the test medium/microbial strains, and this is probably the main reason for variability in results from different centers. Treatment of a disease with modern medicine is often and generally associated with the development of side effects. Hence, the use of plant products has been increasing world-wide to lower medicinal side effects The antimicrobial properties of plant extracts are enviable tools for controlling undesirable microorganisms especially in the treatment of many infections. Moderate daily consumption of greentea killed *Staphylococcus aureus, Vibrio parahemolyticus, Clostridium perfringens, Bacillus cereus, Pleisomonas shigelloides*, etc. Green tea contains between 30% and 40%, andblack tea contains between 3% and 10% of water extractablephenols. Recently, researchers investigated minimum inhibitory dose (MID) of greenteas (Sea Dyke, Ito En, and Boh), black teas (Boh CameronHighlands and Boh Bukit Cheeding), and herbal tea (Ho YanHor) extracts against gram-positive, Micrococcus luteus,Staphylococcus aureus, and Bacillus cereus, and gram-negative, *Escherichia coli, Salmonella typhi*, and *Pseudomonas aeruginosa*, using the disc-diffusion method. The author stated that extracts and fractions of all six tea varieties showedno activity against the three gram-negative bacteria. Green teas inhibited all three gram-positive bacteria with *S. aureus* being the least susceptible one. Black and herbal teas inhibited the growth of *M. luteus* and *B. cereus* but not *S. aureus*. Gram-negative bacteria are less susceptible to antibiotics as their outer membrane of lipoproteins and lipopolysaccharides is able to regulate the access of antibacterial agents into the underlying structures. Outer cell membranes of gram-negative organisms have several lipid compounds that protect cells from antimicrobial agents. Varying degrees of sensitivity of different microorganisms may be due to both the intrinsic tolerance of microorganisms and the nature and combinations of phytochemical compounds present in tea extracts. **Tables 2.2.3.1 and 2.2.3.2** divulge the antibacterial and antifungal activities of different tea extracts, respectively. The antimicrobial and microbiocidal activities of tea extracts, (i) epigallocatechin gallateand theaflavin digallate (TF3) against Mycoplasma. Green and black teas presented antimicrobial activities against *M. pneumoniae*. At a concentration of 0.2%, green and black teas confirmed microbicidal activities against *M. pneumoniae* and*M. orale* but not against *M. salivarium*. Extracts of Pu-erh teashowed slight microbicidal activity against *M. pneumoniae* and*M. orale*. EGCG purified from green tea and TF3 from blacktea markedly showed microbicidal activities against *M. pneumonia, M. orale*, and *M. salivarium*. As reviewed, the antimicrobialactivities of tea extracts depend on the variety of evaluated tea.

Table 2.2.3.1:Antibacterial activity of various tea extracts [Md. W. Siddiqui et al 2016]

Type of tea	Effective against	Test method	Extraction/ solvent used
Black Green	*Streptococcus mutans*	Agar-well diffusion method Ethanol	Methanol
Black	*Staphylococcus aureus*, *Salmonella typhi*, *Pseudomonas aeruginosa*	Agar-gel diffusion inhibition test	Water
Black	*Salmonella enteric*, *Vibrio cholera*	Agar-diffusion technique Agar-dilution technique	Ethanol
Green	*Listeria monocytogenes*	Disc-diffusion method Agar-diffusion method Microbroth-dilution techniques	Methanol Water
White	*Staphylococcus aureus*, *Escherichia coli*, *Pseudomonas aeruginosa*	Disk-agar diffusion assay	Water
Fresh leaves Green	*Staphylococcus aureus*, *Listeria monocytogenes* *Hafnia alvei*, *Salmonella enteric*, *Escherichia coli*, *Bacillus cereus*	Disc-diffusion method	Methanol Ethanol Water
Green tea leaves	*Staphylococcus aureus*,	Agar-well diffusion method	Methanol
	Escherichia coli, *Bacillus cereus*, *Pseudomonas aeruginosa*	Paper disc-diffusion method	
Green	*Proteus mirabilis*, *Streptococcus pyogenes*	Paper disc method	Water
Pu-erh tea	*Staphylcoccus aureus*, *Bacillus subtilis*, *Escherchia coli*	—	Water
Black Green Herbal	*Micrococcus luteus* *Staphy lococcus aureus* *Bacillus cereus*, *Escherichia coli* *Salmonella typhi*, *Pseudomonas aeruginosa*	Disc-diffusion method	Hot water
Fresh tea leaves	*Escherichia coli*	Agar-gel diffusion inhibition test, paper disk-diffusion inhibition test	Water
Green tea Black tea	*Enterococcus faecalis* *Salmonella typhi*, *Streptococcus aureus*, *Pseudomonas aeruginosa*, *Vibrio cholerae*		Methanol
Green Black	*Streptococcus mutans*	Agar-dilution method	Methanol
Black	*Staphylococcus aureus*, *Listeria monocytogenes*, *Escherichia coli*, *Hafnia alvei*, *Yersinia enterocolitica*, *Bacillus cereus*	Disc-diffusion method	Methanol Ethanol *N, N*-dimethylformamide Acetone
Green	*Lactobacillus acidophilus*,	Paper disc-diffusion inhibition test	Water
	Streptococcus mutans		Ethanol

Table 2.2.3.2: Antifungal activity of various tea extracts [Md. W. Siddiqui et al 2016]

Type of tea	Effective against	Test method	Extraction/ solvent used
Green	*Candida albicans*	Paper disc method	Water
Green tea leaves	*Aspergillus parasiticus*	Agar-well diffusion method	Methanol
Fresh tea leaves Green tea Black tea	*Aspergillus flavus* *Fusarium sp.* *Aspergillus fumigates* *Aspergillus niger* *Candida albicans*	Paper disc-diffusion method Agar-gel diffusion inhibition test Paper disk-diffusion inhibition test	Water Methanol

2.2.3.4.1 The antibacterial activity of TTO Banes-Marshall et al (**2001**) reported that TTO is bactericidal in nature, although it may be bacteriostaticat lower concentrations. The activity of TTO against antibiotic-resistant bacteria has attracted considerable interest, especially with methicillin-resistant *S. aureus* (MRSA) receiving the most attention so far. In a previous study, Inouye et al (**2001**) reported that vaporized TTO can inhibit several bacteria spp. including; *Mycobacterium avium* ATCC 4676, *E. coli, Haemophilus influenzae, Streptococcus pyogenes* and*S. pneumoniae*. Treatment of *S. aureus* with TTO resulted in the leakage of K ions (**Cox et al 2000, Hada et al 2003**), and inhibited its respiration (**Cox et al 2000**). Reichling et al (**2002**) stated that treatment with TTO sensitized*S. aureus* cells to sodium chloride, and caused morphological changes. Similarly, Noormandi and Dabaghzadeh(**2015**) reported that the polyphenolic components of green tea have antibacterial potential.

2.2.3.4.2 The antifungal potential of TTO Several years before, Senthaamarai et al (**2015**) pointed that the germinated fungal conidia are significantly more susceptible to TTO than non-germinated ones, suggesting that the intact conidial wall confers considerable protection. In addition, TTO vapors have also been reported to inhibit fungal growth and affect their sporulation (**Schnitzler et al 2001**), and they also inhibit the formation of germ tubes or mycelial conversion in *C. albicans* (**Schnitzler et al 2001**).Noumi et al (**2011**) reported that at low concentrations as 0.3125mg/ml of tea tree essential oil and 0.078mg/ml of eucalyptus oil, they drastically impair the maximum yield and growth rate of *C. albicans* and *C. glabrata*, suggesting that essential oils can be used as part of the efficient oral hygiene regime.

2.2.3.4.3 The antiviral efficacy of TTO Previously, Schnitzler et al (**2001**) documented that at high concentration of TTO (0.003%), it reduced Herpes Simplex Virus (HSV-1) titers by 98.2%, and HSV-2 titers by 93.0%. They also reported that by applying TTO at different stages of the virus replicative cycle, it had the greatest effect on the intact virus (prior to infection of cells); however, when it was applied during the adsorption period, a slight reduction in the plaque formation was observed. Another study by Mikus et al (**2000**) evaluated the activities of 12 different essential oils including TTO for their activities against HSV-1 in Vero cells, where TTO exerted most of its antiviral activity on the free virus. They observed that 1% of the essential oils inhibited plaque formation completely, while 0.1% TTO reduced plaque formation by approximately 10%.

2.2.4.5 Antioxidant and antimicrobial activities of Tea Compared with other types of tea, green tea is undoubtedly oneof the most potent natural preservatives applied to many foods as antioxidant and antimicrobial agents (**Zhu et al 2005,Almajano et al 2008, Su et al 2008, Kristanti and Punbusayakul 2009**). Several previous studies of Sakanaka et al (**2000**), Wang et al (**2000**) have demonstrated the antimicrobial efficacy of green tea extract in vitro against some potent pathogens such as;*Shigella dysenteriae,Salmonella* sp.,*E. coli, S. aureus* and *Listeria monocytogenes*.The antioxidant activity of green tea extract has also been verified in raw, frozen and cooked meat patties (**Jo et al 2003, Mitsumoto et al 2005**), and in sausage (**Bozkurt 2006**). Previously, Kristanti et al (**2014**) investigated the effects of two brands of commercial Assam green tea infusion on the microbial growth, anti-lipid oxidation and color change in cooked beef, and recorded a significant reduction in the populations of *Staph. aureus, Listeria monocytogenes, S. typhimurium* and *E. coli*, to an undetectable level in the cooked beef within 2days of storage at 4°C. In previous studies, Pokorný et al (**2001**), Michalczyk and Zawislak (**2008**) pointed that besides acting as antimicrobial agents, the polyphenoliccompounds in green tea are primarily known as antioxidant agents that can retard the free radical chain reaction during the process of oxidation. Dorota (**2014**) studied the effect of addition of green tea on the selected properties of probiotic milk, and reported that green tea can be successfully employed as a functional supplement for probiotic milk, as it added an extra value to the known health benefits of the probiotics. Moreover, Maksum et al (**2013**) concluded from thei study that*C. sinensis* leaves extract could be useful in combating emerging drug-resistance caused by MRSA and *P. aeruginosa*.

2.2.3.6 Antibacterial activity of differenttea extracts [Siddiqui et al 2016]

2.2.3.6.1 Black Tea Extracts The antibacterial potential of themethanolic extracts of *Camellia japonica* against foodborne pathogens. The extract showed good bactericidalresponse against pathogens *Salmonella typhimurium, E. coli, Listeria monocytogenes,* and *Staphylococcus aureus.*

While observing antibacterial activity against *Streptococcus mutans*, demonstrated that the crudemethanolic extract of black tea showed full growth of bacteria between the concentrations of 0.1 and 0.5mg/mL, while the moderate activity was recorded at the concentrations of 0.6–1.0mg/mL. The crude methanolic extract of black tea, which was diluted in ethanol, showed good bactericidal activity nearly at all concentrations. The antibacterial activity of the extracts of mint, tea, and mint tea (1:1) against microorganisms such as *S. aureus, S. typhi*, and *P. aeruginosa*. Tea extract showed strong antibacterial activity against *S. aureus* with minimum inhibitory concentration (MIC) at 150L where the zone of inhibition was 27mm. Tea extracts of 200L and 250L concentrations produced a significant antibacterial effect on *S. typhi* and *P. aeruginosa*, and the zone of inhibition was observed at 24.1 and 18.1mm, respectively. The methanolic extracts of green and black teas were examined on *Streptococcus mutans* (a dental caries causing bacteria) and five different concentrations (50,100, 200, 300, and 400 mg/mL) were used. Black tea had an antibacterial effect on 100 to 400 mg/mL concentrations. The MIC of black tea was 50 mg/mL. The mean zone diameter of inhibition (ZDI) was 10.9 mm for methanolic extract of black tea. Typhoid and cholera caused due to the infection of *Salmonella enterica* serovar *Typhi* and *Vibrio cholerae*, respectively, are endemic in many developing countries such as India. Tea extracts have an influence on these microorganisms. It has been reported that antibacterial activity of *C. sinensis* against different bacterial genera, including *S. typhi*. Extraction solvents (acetone, N,N-dimethyl-formamide (DMF), ethanol, and methanol) used as controls had no inhibitory effects on the six bacteria. All black tea extracts except the 50% methanol showed varying antibacterial activity against *S. aureus, H. alvei*, and *B. cereus*, depending on the solvent used, but no activity was shown against other bacteria. Gram-positive bacteria were more sensitive to tea extracts than gram-negative ones. Recently, it has been determined ZDI and MIC of black tea extracts against *Salmonella enterica* serovar *Typhi*and *Vibrio cholerae*. They reported that *S. typhi* isolates had ZDI of 12–17mm, and it ranged from 13–21 mm for *V. cholerae Ogawa* isolates, and thus all bacterial isolates were sensitive to tea extracts. All *S. typhi* isolates (12; 100%) showed extract MIC of 400–600mg/mL, and most of them (7; 58.33%) had MIC of 500mg/mL. Among nine *V. cholera Ogawa* isolates, six (66.67%) had MIC of 400–600mg/mL, and the rest three (33.33%) isolates had 200–300mg/mL. The extract at 32mg/mL started to exhibit killing effect on *S. typhi*, while the extract had a similar effect on *V. cholerae Ogawa* at 64 m g/mL. The extract showed bactericidal activity at the concentrations of 512mg/mL and 256mg/mL against *S. typhi* and *V. cholerae Ogawa*, respectively, leaving 2.19 log 10 cfu/mL and 2.58 log 10cfu/mL, respectively, after 24h. The black tea extract had the strongest inhibition activity compared with other teas (Puerh and green tea) against human intestinal bacteria, and *S. enteritidis* was inhibited largely by black tea than by green tea.

2.2.3.6.2 Green Tea Extracts The effectiveness of green tea as an antimicrobial agent isattributed to its low degree of fermentation. During the fermentation process, catechins, such as EGCG, are destroyed, reducing the tea's antimicrobial properties. Phenolics of green tea areconsidered to scavenge free radicals and to chelate metal ions,and being used as natural antioxidants, antimicrobial, and anti-viral agents. Phenolics extracted from green tea extract have inhibitoryeffects on Gram-positive and Gram-negative bacteria. Green teacatechins reduce the heat resistance of spore-forming thermophilic spoilage bacteria *Bacillus stearothermophilus* and *Clostridium hermoaceticum* that proliferate in vending machines,causing sour spoilage of milk and other drinks. Use of phenolics or other natural antimicrobials mayresult in the reduction of temperature used in the thermal processing of foods because widely consumed well-done meat products exposed to high cooking temperatures employed to killpathogens also induce the formation of potentially carcinogenicheterocyclic amines.Antimicrobial activity ofgreen tea and attributed such action to the presence of catechins at 1.5% concentration. Water extracts of different teas, including green tea, had antibacterial activity against *S. aureus* and *B. subtilis* at a finalconcentration of 2mg/mL but no activity against *E. coli.* Dry tea flush and greentea, in particular, are capable of killing bacteria. Dry teaproved most effective against *B. subtilis, Proteus vulgaris*, and*Staphylococcus aureus*, but green tea demonstrated better performance against *E. coli* and *Salmonella*. The antioxidant and antibacterialactivities of methanol, ethanol, and water extracts and theircrude, ethyl acetate and water fractions derived from fresh tealeaves and green tea against a number of bacteria species. Inboth fresh tea leaves and green tea, ethyl acetate fractions andthe crude extracts were found to have activity against only twobacteria, namely *S. aureus* and *B. cereus*, depending on theextract type. In fresh tea leaves, *S. aureus* was inhibited byboth ethyl acetate fractions and crude extracts but *B. cereus*was inhibited by only ethyl acetate fractions. The strongerantibacterial activity of fresh tea leave extracts (as comparedwith green tea extracts) is related to phenolic contents.Antibacterialactivity by measuring ZDI and MIC at four different concentrations (10, 25, 25, and 100mg/mL) of the methanolic extractof green tea leaves. Fifty percent methanolic extract of leavesshowed greater antibacterial activity against *B. cereus* followed by *P. aeruginosa*. The ZDI was found to increase as theconcentration of the extract is increased. The MIC was100mg/mL for *E. coli* and 10mg/mL for *B. cereus*. In assessing the antibacterial activity of EGCG against various strainsof *Staphylococcus* (Gram-positive cocci) and gram-negativerods, including *E. coli, Klebsiella pneumonia*, and *Salmonella*,it was revealed that 50–100mg/mL was required to inhibit thegrowth of *Staphylococcus* whereas the concentration higherthan 800mg/mL was required to inhibit Gram-negative rods.The antibacterial activityof aqueous and ethanolic extract of green tea using paper diffusion test and MIC. Water extract of *C. sinensis* produceszones of inhibition of 6–18 mm and 8–27 mm against *L. acidophilus* and *S. mutans*, respectively. Ethanolic extracts of *C.sinensis* produced larger zones of inhibition, 15–33mm and19–35mm against *L. acidophilus* and *S. mutans*, respectively.The MIC for aqueous extract against *L. acidophilus* and *S.mutans* was 0.9 and 0.8mg/mL, respectively, while for ethanolic extract, it was 0.7 and 0.7mg/mL against *L. acidophilus*and *S. mutans*, respectively.Greater antibacterialactivity in fresh green tea extracts (methanolic and aqueous)against *E. coli, Enterococcus faecalis, Staphylococcus aureus*,and *Pseudomonas aeruginosa*. The fresh green tea extractwith methanol was found to have high antimicrobial activityfollowed by commercial green tea leaves, and the least activitywas found in dust tea. The MIC of black tea is effective *againstStaphylococcus aureus, Salmonella typhi*, and *E. coli* that werefound to be at higher concentrations when compared with freshgreen tea leaves.The antimicrobial activity of Iranian green and black teatested against *Streptococcus mutans* revealed that green teahad an antibacterial effect at 100 to 400mg/mL concentrations. The MIC of green tea was 150mg/mL, and the mean ZDI was 9.5mm for methanolic extract of green tea. Antibacterialactivity of methanolic extract of green tea against *Listeria monocytogenes* and showed that the methanolic extracthad greater antibacterial property (ZDI D 20.1mm) in relationto hydrophilic extract (ZDI D 10mm). The water extract didnot show inhibitory effect, while the methanolic extract hadZDI of 15 mm. The MIC for methanol and aqueous leafextracts was 0.26 and 0.68 mg/mL, respectively (**Siddiqui et al 2016**).

2.2.3.6.3 White Tea Extracts Compared with black and green teas, only few studies havebeen carried out on the antimicrobial activity of white tea.White tea differs from others regarding the processing methods. This tea is obtained from young leaves, even before thebuds are completely opened. Recently,the antimicrobial activity of aqueous extract ofwhite tea against different bacteria. Antimicrobial activity wasverified only in samples at 5 to 20mg/mL concentrations, inwhich ZDI was observed against *S. aureus*, with no significantdifferences among them. Similarly, the MIC was observed to be 10mg/mL for this type of microorganism (**Siddiqui et al 2016**).

2.2.3.7 Antifungal activity of tea extracts Different tea compounds and tea extracts exhibited variabletime- and concentration-dependent fungicidal activitiesagainst several fungi. Antifungal activity against two*Aspergillus* spp. by measuring ZDI and MIC at four differentconcentrations (10, 25, 25, and 100mg/mL) of the methanolicextract of green tea leaves. They reported no activity of boththe species at any concentration. It reported antifungal activity (AFA) of different teaextracts (fresh leaves, green and black teas) against fungi suchas *Fusarium* sp. *Aspergillus fumigates, Aspergillus niger*, and*Candida albicans*. With black tea extract, *A. niger*did not show any positive results. *C. albicans* found tobe sensitive to fresh green tea extract. The fresh green teaextract with methanol had higher antimicrobial activity thanaqueous extract. The effectiveness of green tea extracts against *C. albicans* but only when the pH was adjusted to 6.5 (**Siddiqui et al 2016**).

2.2.3.8 Conclusion

Different tea products have different biochemical profiles, and are great sources of antioxidants. Tea has inhibitory activities on carcinogenic and periodontopathic bacteria. Therefore, it can be inferred that the combined use of tea and antibiotics could be useful in fighting the emerging drug resistance problem. However, there are many improvements and experiments that need to be carried out in the future to determine the effectiveness of tea products.

Different in vitro studies provide evidence that tea extracts are potentially a rich source of antimicrobial agents against several microorganisms with the bioactivity attributed by their phytochemical constituents and mineral contents. The antimicrobial activity may also be attributed to a synergistic action of phytotherapeutic composition or secondary metabolites. All tea extracts (black, white, and green) showed significant antimicrobial effects along with other beneficiary effects such as free radical scavenging property and anti-atherogenic property. The current review furnishes credence to their ethnic-pharmacological use as a remedy to treat infections and diseases caused by microorganism. The analyzed information suggests that the combined use of catechins and antifungal drugs may be useful in the treatment of super-infections of oral cavities, intestines, and vagina resulting from over consumption of antibiotics. The available information provides a basis for future studies on the potential of tea compounds to inactivate bacteria in liquid and solid foods and to protect humans against infectious diseases. Precise screening of compounds may lead to the identification of contents that are sufficiently potent to be useful as antibacterial, antifungal, or antiviral chemotherapeutics. In addition to the structural alteration of weak and moderately active antimicrobial flavonoids, investigation into the mechanisms of action of these compounds is likely to be a productive area of research. Such information may assist in the optimization of a lead compound's activity, provide focus for toxicologicalattention, and aid in the anticipation of resistance. As future directions, tea extracts must be explored as a natural source of antimicrobial compounds with potential use as food additives. The applications of tea extracts may be contemplated to extend the shelf life of foods that are decomposed by microorganisms; however, the collateral effects that its addition may cause on the quality of the treated food must also be considered.Some of the collateral effects could be noticed in the flavor andcolor, and therefore on the consumer acceptability of the treated food. Moreover, the possible generation of functional food when the addition of tea extracts occurs to food matrices could be an option. For instance, green tea infusion could be a potential natural preservative that could be used to extend the shelf life of fruit juices by improving their microbiological safety. The applied aspects of tea extract treatments could be considered to optimize the antimicrobial efficacy of extracts, and offer products to consumers treated with natural preservatives.

2.2.3.9 References

Aaron B, Ryan L, Deirdre G, Paddy K. and Michael T. 2006. In vitro activity of tea-tree oil against clinical skin isolates of meticillin resistant and sensitive Staphylococcus aureus and coagulase-negative staphylococci growing planktonically and as biofilms. Journal of Medical Microbiology. 55: 1375-1380.

Agnieszka K and Wilfried A. 2014. Antioxidant Capacity of Tea: Effect of Processing and Storage (Chapter 12); In: Processing and impact on antioxidants in beverages, Academic Press, pp. 109-120.

Aleksandra N, Mohd WR, Tatiana K and Matthias R. 2016. Anticancer Efficacy of Polyphenols and Their Combinations. Nutrients. 8(9): 552.

Almajano MP, Carb, R, Jimenez JAL and Gordon MH. 2008. Antioxidant and antimicrobial activities of tea infusion. Food Chemistry. 108(1): 55-63.

Anton R and Sheila W. 2003. Antioxidant Effects of Tea: Evidence from Human Clinical Trials. The Journal of Nutrition. 133(10): 3285S–3292S.

Arina N, Dedi SD, Betty N, Dwi DR and Wahyu W. 2017. Anti-inflammatory properties of oolong tea (Camellia sinensis) ethanol extract and epigallocatechin gallate in LPS-induced RAW 264.7 cell. Asian Pacific Journal of Tropical Biomedicine. 7(11): 1005-1009.

Banes-Marshall L, Cawley P and Phillips CA. 2001. In vitro activity of Melaleuca alternifolia (tea tree) oil against bacterial and Candida spp. isolates from clinical specimens. Brazilian Journal of Biomedical Science. 58: 139-145.

BlancoAR, Sudano-Roccaro A, Spoto GC, Nostro A and Rusciano D. 2005. Epigallocatechin gallate inhibits biofilm formation by ocular staphylococcal isolates. Antimicrobial Agents Chemotherapy. 49: 4339-4343.

Bozkurt H. 2006. Utilization of natural antioxidants green tea extract and Thymbra spicata oil in Turkish dry-fermented sausage. Meat Science. 73: 442-450.

Carson CF, Hammer KA and Riley TV. 2006. Melaleuca alternifolia (Tea Tree) Oil: a Review of Antimicrobial and Other Medicinal Properties. Clinical Microbiology Reviews. 19(1): 50-62.

Chen CN, Lin CPC, Huang KK, Chen WC, Hsieh H.P, Liang PH and Hsu JTA. 2005. Inhibition of SARS-CoV 3C-like protease activity by theaflavin-3, 3'-digallate (TF3). Evidence Based Complement Alternativ Medicine. 2: 209-215.

Cho YS, Schiller NL and Oh KH. 2008. Antibacterial effects of green tea polyphenols on clinical isolates of methicillin-resistant Staphylococcus aureus. Current Microbiology. 57: 542-546.

Chung JH, Han JH, Hwang EJ, Seo, JY, Cho KH and Kim KH. 2003. Dual mechanisms of green tea extract (EGCG)-induced cell survival in human epidermal keratinocytes. FASEB Journal. 17(13): 1913-1915.

CongL, Cao C, Cheng Y and Qin XY. 2016. Green tea polyphenols attenuated glutamate excitotoxicity via antioxidative and antiapoptotic pathway in the primary cultured cortical neurons. Oxidative medicine and cellular longevity. 4: 1-8.

Cox SD, Mann CM, Markham JL, Bell HC, Gustafson JE, WarmingtonJR and Wyllie SG. 2000. The mode of antimicrobial action of the essential oil of Melaleuca alternifolia (tea tree oil). Journal of Applied Microbiology. 88: 170-175.

Dorota NL. 2014. Effect of green tea supplementation on the microbiological, antioxidantand sensory properties of probiotic milks. Dairy Science and Technology. 94: 327-339.

Duh PD, Yen GC, Yen WJ, Wang BS and Chang LW. 2004. Effects of Pu-erh tea on oxidative damage and nitric oxide scavenging. Journal of Agricultural Food Chemistry. 52: 8169-8176.

Hada T, Inoue Y, Shiraishi A and Hamashima H. 2003. Leakage of K ions from Staphylococcus aureus in response to tea tree oil. Journal of Microbiological Methods. 53: 309-312.

Hammer KA, Dry L, Johnson M, Michalak EM, Carson CF and Riley TV. 2003. Susceptibility of oral bacteria to Melaleuca alternifolia (tea tree) oil in vitro. Oral Microbiology and Immunology. 18: 389-392.

Hammer KA, Carson CF and Riley TV. 2000. Melaleuca alternifolia (tea tree) oil inhibits germ tube formation by Candida albicans. Medical Mycology. 38(5): 355-562.

Hammer KA, Carson CF and Riley TV. 1996. Susceptibility of transient and commensal skin flora to the essential oil of Melaleuca alternifolia (tea tree oil). American Journal of Infectious Control. 24(3): 186-9.

Hu ZQ, Zhao WH, Asano N, Yoda Y, Hara Y and Shimamura T. 2002. Epigallocatechin gallate synergistically enhances the activity of carbapenems against methicillin-resistant Staphylococcus aureus. Antimicrobial Agents Chemotherapy. 46: 558-560.

Inouye S, Takizawa T and Yamaguchi H. 2001. Antibacterial activity of essential oils and their major constituents against respiratory tract pathogens by gaseous contact. Journal of Antimicrobial Chemotherapy. 47: 565-573.

Janecki A and Kolodziej H. 2010. Anti-adhesive activities of flavan-3-ols and proanthocyanidins inthe interaction of group A-streptococci and humanepithelial cells. Molecules. 15: 7139-7152.

Jie G, Lin Z, Zhang L, Lu H, He P and Zhao B. 2006. Free radical scavenging effect of Pu-erh tea extracts and their protective effect on oxidative damage in human fibroblast cells. Journal of Agricultural Food Chemistry. 54: 8058-8064.

Jo C, Son JH, Son CB and Byun MW. 2003. Functional properties of raw and cooked pork patties with added irradiated, freeze-dried green tea leaf extract powder during storage at 4°C. Meat Science. 64: 17-33.

Kristanti RA, Hadibarata T and Punbusayakul N. 2014. Beneficial effects of commercial Assam green tea infusion on the microbial growth and oxidative stability of cooked beef. International Food Research Journal. 21(4): 1313-1320.

Kristanti RA and Punbusayakul N. 2009. Antioxidant and antimicrobial activity of commercial green tea in Chiang Rai. Acta Horticultural. 837.

Kunle OF, Egharevba HO and Ahmadu PO. 2012. Standardization of herbal medicines-A review. International Journal of Biodiversity and Conservation. 4(3): 101-112.

Kuo KL, Weng MS, Chiang CT, Tsai YJ, Lin-Shiau SY and Lin JK. 2005. Comparative studies on the hypolipidemic and growth suppressive effects of oolong, black, Pu-erh, and green tea leaves in rats. Journal of Agricultural Food Chemistry. 53: 480-489.

Kyungseop A. 2017. The worldwide trend of using botanical drugs and strategies for developing global drugs. BMB Reports online. 50(3): 111-116.

Lee JH, Shim JS, Chung MS, Lim ST and Kim KH. 2009. In vitro anti-adhesive activity of green tea extract against pathogen adhesion. Phototherapeutic Research. 23: 460- 466.

Li M, Chen L, Lee DHS, Yu LC and Zhang Y. 2007. The role of intracellular amyloid beta in Alzheimer's disease. Progressive Neurobiology. 83(3): 131-139.

Maksum R, Rafael AA, Berna E and Conny RT. 2013. Antimicrobial activity of green tea extract against isolates of methicillin-resistant Staphylococcus aureus and multi-drug resistant Pseudomonas aeruginosa. Asian Pacific Journal of Tropical Biomedicine. 3(8): 663-667.

Martins E. 2013. The growing use of herbal medicines: issues relating to adverse reactions and challenges in monitoring safety. Frontier in Pharmacology. 4: 177.

Messager S, Hammer KA, Carson CF and Riley TV. 2005. Effectiveness of hand-cleansing formulations containing tea tree oil assessed ex vivo on human skin and in vivo with volunteers using European standard EN 1499. Journal of Hospital Infection. 59: 220-228.

Michalczyk M and Zawislak A. 2008. The effect of tea infusions on the proliferation of selected bacteria important for the human intestinal tract. Acta Scientiarum Polonorum Technologia Alimentaria. 7: 59-65.

Mikus J, Harkenthal M, Steverding D and Reichling J. 2000. In vitro effect of essential oils and isolated mono- and sesquiterpenes on Leishmania major and Trypanosoma brucei. Planta Medica. 66: 366-368.

Mitsumoto M, O-Grady MN, Kerry JP and Joe BD. 2005. Addition of tea catechins and vitamin C on sensory evaluation, colour and lipid stability during chilled storage in cooked or raw beef and chicken patties. Meat Science. 69: 773- 779.

Md. Wasim Siddiqui AB, Sharangi JP, Singh PK, Thakur JF, Ayala-Zavala ASand Dhua RS. 2016. Antimicrobial Properties of Teas and Their Extracts in vitro, Critical Reviews in Food Science and Nutrition, 56:9, 1428-1439, DOI:10.1080/10408398.2013.769932

Napatt K, Pakakrong K, Prin V, Nisakron T, Yingampa G, Piyamitr S and Chagriya K. 2019. Herbal or traditional medicine consumption in a Thai worker population: pattern of use and therapeutic control in chronic diseases. BMC Complement Alternative Medicine. 19: 258.

Noormandi A and Dabaghzadeh F. 2015. Effects of green tea on Escherichia coli as a uropathogen. Journal of Traditional and Complementary Medicine. 5(1): 15-20.

Noumi E, Snoussi M, Hajlaoui H, Trabelsi N, Ksouri R, Valentin E and Bakhrouf A. 2011. Chemical composition, antioxidant and antifungal potential of Melaleuca alternifolia (tea tree) and Eucalyptus globulus essential oils against oral Candida species. Journal of Medicinal Plants Research. 5(17): 4147-4156.

Ohishi T, Goto S, Monira P, Isemura M and Nakamura Y. 2016. Anti-inflammatory Action of Green Tea. Anti-inflammatory Anti-allergy Agents Medicinal Chemistry. 15(2): 74-90.

Orly W, Tamar A, Silvia M and Moussa BHY. 2009. Neuroprotective molecular mechanisms of (-)-epigallocatechin-3-gallate: a reflective outcome of its antioxidant, iron chelating and neuritogenic properties. Genes Nutrition. 4(4): 283-296.

Pokorný J, Trojáková L and Takácsová M. 2001. The use of natural antioxidants in food products of plant origin. Antioxidants in Food. 355-372.

Priyanka C, Sangita C, Protapaditya D and Sanjib B. 2012. Evaluation of anti-inflammatory effects of green tea and black tea: A comparative IN VITRO study. Journal of Advanced Pharmaceutical Technology and Research. 3(2): 136-138.

Qin XY, Cheng Y and Yu LC. 2012. Potential protection of green tea polyphenols against intracellular amyloid beta-induced toxicity onprimary cultured prefrontal cortical neurons of rats. Neuroscience Letters. 513(2): 170-173.

Reichling J, Weseler A, Landvatter U and Saller R. 2002. Bioactive essential oils used in phytomedicine as anti-infective agents: Australian tea tree oil and manuka oil. Acta Phytotherapeutica. 1: 26-32.

Reygaert WC. 2017. An Update on the Health Benefits of Green Tea. Beverages. 3(1): 6.

Sakanaka S, Juneja LR and Taniguchi M. 2000. Antimicrobial effects of green tea polyphenols on thermophilic spore-forming bacteria. Journal of Bioscience and Bioengineering. 90(1): 81-85.

Schnitzler P, Schon K and Reichling J. 2001. Antiviral activity of Australian tea tree oil and eucalyptus oil against Herpes simplex virus in cell culture. Pharmazie. 56: 343-347.

Senthaamarai R, Sharyn G, Michael T and Dino P. 2015. An Evaluation of Antifungal Agents for the Treatment of Fungal Contamination in Indoor Air Environments. International Journal of Environmental Research and Public Health. 12(6): 6319-6332.

Sharma A, Gupta S, Sarethy IP, Dang S and Gabrani R. 2012. Green tea extract: Possible mechanism and antibacterial activity on skin pathogens. Food Chemistry. 135: 672-675.

Stapleton PD, Shah S, Hamilton-Miller JMT, Hara Y, Nagaoka Y and Kumagai A. 2004. Anti-Staphylococcus aureus activity and oxacillin resistance modulating capacity of 3-O-acyl-catechins. International Journal of Antimicrobial Agents. 24: 374-380.

Su P, Henriksson A, Nilsson C and Mitchell H. 2008. Synergistic effect of green tea extract and probiotics on the pathogenic bacteria, Staphylococcus aureus and Streptococcus pyogenes. World Journal of Microbiology and Biotechnology. 24(9): 837-842.

Wang H, Provan GJ and Helliwel K. 2000. Tea flavanoids: their function, utilisation and analysis. Science and Technology. 11: 152-160.

Zhao WH, Hu ZQ, Hara Y and Shimamura T. 2002. Inhibition of penicillinase by epigallocatechin gallate resulting in restoration of antibacterial activity of penicillin against penicillinase-producing Staphylococcus aureus. Antimicrobial Agents Chemotherapy. 46: 2266-2268.

Zhao WH, Hu ZQ, Okubo S, Hara Y and Shimamura T. 2001. Mechanism of synergy between epigallocatechin gallate and beta-lactams against methicillin-resistant Staphylococcus aureus. Antimicrobial Agents Chemotherapy. 45(6): 1737-1742.

Zhu YX, HuangH and Tu YY. 2005. A review of recent studies in China on the possible beneficial health effects of tea. International Journal of Food Science and Technology 41:333-340.

2.2.4 JUTE [Md. Mahabub et al 2019, Aleya et al 2021, Zakaria and Shuranjan 2022, Taslima et al 2022, Aleya et al 2022]

2.2.4.1 Potential Healing Powers of Jute [Md. Mahabub et al 2019]

2.2.4.1.1 Introduction

Jute is cultivated to provide bark for the production of fibers. Its mucilaginous leaves are used in food as a vegetable. It is grown for both fiber and culinary purposes [**1**]. The jute plant contains p-coumaric acid, ferulic, caffeic, vanillic, p-hydroybenzoic, protocatechuic, vanillic acids, β–sitosterol in 80% aqueous ethanol extract of *C. capsularis* [**2-3**]. Demand for medicinal plants is increasing in both usually developed and developing countries due to growing recognition of natural products being equally effective, safe, non-nacotic, affordable and having no side effects. Higher plants represent a rich source of new molecules with pharmacological properties, which are lead compounds for the development of new drugs. Natural products have been an integral part of the ancient traditional medicine system, e.g. Chinese, Ayurvedic and Egyptian [**4**]. Extracts from natural products containing medicinal, cosmetic, aromatic, dyes or pharmaceutical properties have been studied in many researches. Several researchers have shown that the ingestion of natural extracts containing antioxidants, especially polyphenolic compounds, is associated with lower rates in coronary heart disease, cancer and diabetes [**5-6**]. The antioxidant potential of polyphones is based on their ability to capture and react with free radicals. Polyphenols are the most abundant secondary metabolites found in plants and include several classes of compounds such as phenolic acid, colorfull anthocyanins, simple and complex flavonoids. In this review, a compilation of literature is made in order to produce a comprehensive report related to phytochemistry of different plant parts of jute and various biological properties exhibited by purified compounds as well as by the crude extracts.

2.2.4.1.2Common names of jute [7] Bengali: Paat sag, Tita pat (White jute), Misti pat (Tossa) Nalita, Nalta, Mara Sag; Arabic: Joot abiadh, Melukhiya; Chinese: Huang ma; English: Jute; French: Chanvre de Calcutta; German: Rundkapsel jute; Hindi: Patta shaak, Italian: Juta; Kenya: Mrenda, Murere; Korean: Hwangma; Nigeria: Rama; North Africa: Melukhiya; North Sudan: Khudra; Phillipines: Saluyot; Portuguese: Juta; Seria Leone: Krain; Spanish: Yute blanco; Swedish: Jute and Tamil: Pirattai keetai.

2.2.4.1.3Jute components

2.2.4.1.3.1 Jute leaf Plants generally produce many secondary metabolites which constitute an important source of many pharmaceutical drugs. The unique chemistry of jute is highly complex. It contains hundreds of different beneficial compounds known as phyto-chemicals. Jute leaf is a unique plant part which is a rich source of many chemical compounds and plays an important role in human body. The crude methanolic extract of jute leaves extract contained the highest amount of polyphenols. It contained also cardiac glycosides, triterpenoids, ionones, flavonoids, coumarins, steroids and many other secondary metabolites. Nutritional chemicals of each 100g of the leaves contained 43–58 calories, 4.5–5.6g protein, 0.3g fat, 7.6–12.4g total carbohydrate, 80.4–84.1g water, 1.7-2.0g fibre, 2.4g ash, 266-366mg,Ca, 97–122mg, P, 7.2-7.7mg Fe, 12mg Na, 444mg K, 6, 410-7, 850μg beta carotene equivalent, 0.13–0.15mg thiamine (Vitamin B1), 0.26–0.53 mg riboflavin (Vitamin B2), 1.1–1.2mg niacin and 53-80 mg ascorbic acid (Vitamin C) [**8-9,1**]. Folic acid (folate, B vitamin) substantially higher (0.03mg/L) than that of other folacin rich vegetables and iron 72 μg/g. Jute leaves yielded a lot of mucilaginous polysaccharide (acidic polysaccharide) and anthocyanin. Leaf extracts yielded alkaloids, terpenoids, tannins, flavonoids, glycosides [**10**]. It contains two functional compounds, phytol and monogalactosyl diacylglycerol that protect from free radicals [**11**]. Capsin and capsugenin isolated from the leaves of*C. capsularis* and capsin responsible for the major bitter taste [**12**]. β-sitosterol, scopoletin and fusidic acid also isolated from the leaves [**13**]. Caffeine and catechine isolated from *C. capsularis* leaves extract [**14**]. Oleanolic acid was isolated from the leaves of *C. olitorius* of Egyptian origin. Four triterpenoid glycosides (chorchorusins A,B, C and D) isolated from aerial parts of *C. acutangulus*. Leaves of *C. olitorius* contain ionone glucosides which are corchoionosides A, B and C, an ionone glucoside (roseoside) and a monoterpene glucoside betulabuside A. Leaves of*C. olitorius* gave four higher fatty acids with a trienone system, corchorifatty acids A, B, C, D, an uncanoic acid, corchorifatty acid E and a trihydroxy fatty acid, corrchorifatty acid F. Ali and his colleagues [**15**] evaluated the biochemical components of different varieties jute leaves as affected by four months' storage after direct sundry and hot water treatments (HWT) applied. The results revealed that the products produced by the HWT better than direct sundry treatment, which were similar moisture, protein, fat, fibre, mineral and pigment factor similar to fresh condition. HWT dried leaves showed better results in term of degree of acceptability though sensory evaluation. Ali and his colleagues [**16**] to study of storage properties of jute leaves by blanching method. Jute leaf was preserved by three different treatments and compared with refrigerator. High density polythene is the best alternative to refrigerator. It preserves 2.5, 3, and 2.5 times nearly beta carotene, ascorbic acid and total chlorophyll respectively better than control condition. In Consider the vitamins, minerals and other phyto-chemicals, the leaf of jute plant of is a power house of nutrients.

2.2.4.1.3.2 Jute seeds The seeds are poisonous to mammals and insects. The seeds contain cardiac glycosides, corchorin, corchortoin, corchoroside A and B, biosides, olitoriside, erysimoside, strophantidol glycosides, oligosaccharide and olitoriside. It also contains 2.25% raffinose, 11.3–14.8 oil (16.9% palmitic acid, 3.7% stearic acid, 62.5 linoleic acid, 0.9% linolenic acids, 1.8% behenic acid, 1.1 % lignoceiic acid and 9.1% oleic acid) and large portions of B, Mn, Mo and Zn [**17-18**]. Active principle of the jute seed is chorchorin, a glycoside ten times bitter than quinine sulfate.

2.2.4.1.3.3 Jute stem (bark/ fibre) The polysaccharides and lignin are the major constituents in bark, stem and fibre. Glucose, fructose, sucrose, six low molecular weight sugar alchohols and two inositols were identified and quantified in bark. It contained many free glycosidic and ester –linked phenolic acid. The phenolic acid in fresh bark contained ferulic and p-coumarin acids. Jute fibers collect from the bast or outer region of the stem after biological retting of the whole plant. The lipid and lignin composition of jute fibers has been characterized. The most predominant lipophilic compounds were high molecular weight ester waxes (24% of total extrat), followed by free fatty acids (17%), free fatty alcohols (17%) and α–hydroxyl fatty acids (14%). Additionally, significant amounts of alkanes (6%), ω–hydroxyl fatty acids (6%), sterols (6%), steroid and triterpenoids ketones (3%) and steryl glycosides (1%)were also identified. The main inter–unit linkage present in, was the β–4 aryl–ether bond (72%) followed by β–β resinol–type substructures andwith lower amounts of β–5 phenylcoumaran and β–1 spirodienone substructures [**19**].

2.2.4.1.3.4 Jute root Cardiac glycosides (corchoroside–A and cannogenol), steroids (β–sitosterol andstigmasterol 3-O-β-Dglucoside), flavonoids (quercetin), terpenoids (betulinic and oleanolic acid), corosin were isolated from root extract of *C. capsularis* [**13**]. It also contains 0.2% corosin. Ursolic acid, corosolic acid and oxo-corosin were isolated from fresh (non-dried) roots of *C. capsularis* and *C.olitorius*[**20**].

2.2.4.1.4Properties of jute Jute leaves considered carminative, laxative, febrifuge, purgative, stimulant demulcent, tonic, demulcent, appetizer and stomachic. The seed corchorin considered toxic and poisonous; some studies suggest a digitalis–effect on the heart. Jute fibre is long, soft, shiny, 1-4 meters in length, with a diameter of 17 to 20 microns, with high insulating and anti-static properties, with moderate moisture regain and low thermal conductivity.

2.2.4.1.4.1 Edible use as food In many part of the world, the jute plant has become a vegetable source, In Bangladesh, in spring, shorts are pulled from the fields and eaten with rice asa vegetable. It is a favorite food during the summer months, especially in North and East part of Bangladesh. The people use it to produce mucilaginous soup or sauce, stews, curries vegetable dishes or boil the leaves and others [**7**]. Ali and Nasreen [**21**] study in direct sun dry freeze dry and blanch dried jute leaf in order to assess their potentiality for food product development as vegetables and herbal tea. The blanched dried leaf was the most acceptable product with a ranking with of 'like very much'. n the other hand, four category of tea justified. Sensory evaluation of dried jute tea was third position in terms of colour, flavor and texture, and over acceptability. In the Philippines, *C. olitorius* is known as saluyot, is commonly consumed as a leafy vegetable together with bamboo shoots. In Thai cuisine, the leaves locally known as bai po; are eaten blanched, together with plain rice. In North Africa and the Middle East, the young leaves are known in Arabic as malukhiyah and used as green leafy vegetables. Malukhiyah is eaten widely in Egypt and some consider it the Egyptian national dish. It is featured in cuisines from Labanon, Palestine, Syria, Jordan and Tunisia. In Turkey and Cyprus, the plant is known as molokhiya or molocha is usually cooked into a kind of chicken stew. It has been a staple Egyptian food since the time of the Pharaohs and it is form there that it gains its recognition and popularity. Varieties of mallow leaves stew rice is a well known Middle Eastern cuisine. In Japan dry leaf is used as substitute of coffee and tea. In Europe, jute leaves are used for preparing soup.

2.2.4.1.4.2 Ethnomedicinal (Folklore) properties Jute plant use in ethnobotanical medicine as whole plants, especially the leaves and seeds. The leaf is use to treat a wide range of physical aliment. Some physical ailments that could be treated using jute are laxative, headache, chicken fox, influenza and die worm. If we eat hot day, it reduces the body temperature [**7**]. Corchorus species used as herbal medicine to control or prevent dysentery, worm and conspirator. It has traditional uses for the treatment of fever, chronic cystitis, aches, pains, dysentery, enteritis and pectoral pains [**22**]. Leaves are used medicinally as demulcent, bitter tonic, stomachic, laxative, carminative anthalmintic, astringent and intestinal antiseptic. Infusion of dried leaves bitter and commonly used as a stomachic tonic also used in fever, bilious trouble dysentery, liver disorders, intestinal colic, gastric catarrh, skin diseases, atonic dyspepsia, mild jaundice and other disorders of digestive system. For the treatment of dysentery, dried leaves (in Bangladesh, according to unpublished, suttani) are eaten with rice. It is also used as a paste over swellings and abscesses. In India, leaves were used in stomachic, as carminative, diuretic and for the treatment of dysentery (dried leaves), while, seeds were used as purgative. Ayurvedics usedthe leaves for ascites, pain, piles and tumors. It is also used for cystitis, dysuria, fever and gonorrhea. The cold infusion was said to restore the appetite and strength. The aqueous/alcoholic extracts (containing polysaccharides and oligosaccharide) are used in preparations of skin cosmetics or hair preparations for their moisturizing effect. Hidus reduce the plant to ashes, mix it with honey and use it for obstruction of the abdominal viscera. In south India, the dried plant is used as demulcent. Powder of leaves, 5-10 grains, mixed with powdered turmeric in equal parts, used for dysentery. In India infusion of leaves used as tonic and febrifuge. The tribal people of Bolangir use the plant as remedy against threat miscarriage: equal quantities of tender leaves of *C. olitorius* and Carica papaya are cooked to boiling point, cooled, then used one tumbler 3x daily. It reduces the appearance of the wrinkles and fine lines in the face and body. Research reveals that it was the source of health and beauty of the Egyptian royalties, including Cleopatra. Regular consumption of saluyot can make you appear young [**7**]. Powdered seeds with honey and ginger are good for diarrhea. Grains of the powder mixed with equal amounts of Curcuma longa used for acute dysentery.

2.2.4.1.4.1 Pharmacological effects [Md. Mahabub et al 2019, Zakaria and Shuranjan 2022]

2.2.4.1.4.1.1 Acidic Polysaccharide Kimiko and his colleagues [**23**] isolated acidic polysaccharide in the water soluble mucilage extracted from dried *C. olitorius* leaves. The polysaccharides are rich in uronic acid (65%), and consisted of rhamnose, glucose, galacturonic acid, and glucuronic acid. This polysaccharide mainly consisted of O–4 substituted galacturonic acid and glucuronic acid, and O-2 substituted rhamnoseresidues and most of the (1±4) linked uronic acid residues are substituted at the O-3 position with glucuronic acid residues. This polysaccharide shows proliferative activity toward the murine splenocyte.

2.2.4.1.4.1.2 Anticancer effect Recent epidemiological investigations have been providing increasing evidence that high consumption plans leaves and fruits could be associated withreduced risk of cancer, especially that in the gastrointestinal tract. These epidemiological predictions suggest that plants contain anti-cancer or antitumor promoting components. Toshio and his colleagues [**11**] identification of two antitumor promoter compounds from jute leaves, which are phytol and mono-galactosyldiacylglycerol. The antitumor promoting activity is examined by an immunoblottinganalysis. The detectable amount of each active components increased by treatment of the leaves with hot water. These findings suggest that treatment of vegetables with hot water effectively increased the amount of active compounds with activity against tumor promoting chemicals that may be consumed with food [**1**]. Özlem and his colleagues [**24**] demonstrated the in vitro cytotoxic effect of C. olitorius leaves extract (LE) and seed extract (SE) on ARH-77 cells, with relation to total phenol content (TPC) and free radical scavenging activity (FRSA) of the extracts. The LE had high TPC and FRSA, and cytotoxic inhibitory concentration 50 (IC 50) on ARH-77 cells above its IC 50 for FRSA. The LEhad significantly higher total phenol 78 mg gallic acid equivalents (GAE) (/g extract) than the SE (2 mg GAE/g extract) with significantly higher FRSA (IC 50 LE: 23 μg/mL and IC 50 SE: 10 401μg/mL). Both LE and SE exerted cytotoxic effects on cells after 48 h. The IC 50 of SE (17μg/mL) was lower than LE (151μg/mL), which demonstrates its higher cytotoxicity on cells. The extracts were applied at 150 and 75μg/mL for LE and at 17 and 8.5 μg/mL for SE, and the results of the comet assay revealed that the extracts induced genotoxic damage on ARH-77 cells. In both 48 h leaf and seed extract treatments, genotoxic damage significantly increased with increasing concentrations at relevant cytotoxic concentrations. Mohamed and his colleagues [**25**] isolated and identified from the aqueous extract of *C. olitorius*, which protected cell against cytotoxicity of two mycocotoxins [aflatoxin B_1 (AFB_1) and fumonisin B_1 (FB_1)]. The result showed that both mycotoxins decreased cell viability and increased DNA damage. Cytotoxicity was more pronounced when cells were exposed simultaneously to AFB1 and FB1. So, aqueous extract of *C. olitorius* contains a water-soluble natural chemo-preventative agent for cancer.

2.2.4.1.4.1.3 Antioxidant properties Azuma and his colleagues [**26**] isolation of six phenolic antioxidative compounds [5-caffeoylquinic acid (chlorogenic acid), 3,5–dicaffeoylquinic acid, quercetin 3–galatoside, quercetin 3–glucoside, quercetin 3-(6–malonylglucoside) and quercetin

3-(6- malonylgalactoside) (tentative)] from the leaves of *C. olitorius* by NMR and FAB-MS. The contents of these phenolic compounds, ascorbic acid and alpha tocopherol in determined and their antioxidant activities measurement using the radical generator initiated per oxidation of linoleic acid. Results showed that 5-caffeoylquinic acid was a predominant phenolic antioxidant. An antioxidant is any substance that when present al low concentrations compared with those of anoxidazable substrate significantly delays or prevents oxidation of that substrate. The air dried leaves of Corchorus capsularis (20g) were soaked in distilled water (1:20;$^{w}/_{v}$) for 72h at room temperature. The collected supernatants were tested for the free radical scavenging activity against the DPPH and superoxide anion radical scavenging assays. The extract showed remarkable antioxidant activity in both assays with the percentage of inhibition nearly 90% [**9**].

2.2.4.1.4.1.4 Antinociceptive / anti-inflammatory Zakaria and his colleagues [**27**] and Zainul and his colleagues [**28**] investigated in experimental animal model of *C. capsularis* leaves extrac by chloroform. The antinociceptive activity was measured using the writhing, hot plate and formalin test, while the anti-inflammatory activity was measured using the carrageenan induced paw edema test. The extract, obtain after 72h soaking of the air dried leaves in chloroform followed by a vacuo evaporator to dryness, was weighed and prepared by serial dilution in DMSO in the dose of 20,100,and 200 mg/kg. The extract was administering 30 min prior to subsection to the respective assays. The extract was foundto exhibit significant antinociceptive and anti inflammatory activities. This result confirmed that the traditional claims of using to treat various ailments related to inflammation, fever and painful which supporting its folkloric use.

2.2.4.1.4.1.5 Cardiotonic/Hypertensive effect Several worker'sisolations of corchori, corchorgenin, capsularin, corchoritin, olitoriside, corchosularin and corchortoxinfrom jute seeds. A group of medicines extracted from jute seeds that are called cardio–tonic activity (similar to digitalis genus). It works by inhibiting sodium potassium ATPase. It acts on the contractile force of the cardiac muscle. Because of their potency in disrupting the function of heart, most are extremely toxic. Jute seeds extract showed better activities than corchortoxin. It's corchoroside A and B show digitalis like action [**18-19, 29-31**].

2.2.4.1.4.1.6 Anti–obesity Dietary supplementation with polyphenoliccompounds is associated withreduced diet induced obesity and metabolic disorders in human. The antioxidative properties of polyphenolic compounds contribute to their anti-obesity effect in animal. Wang and his colleagues [**32**] investigate the antiobesity of polyphenolic compounds from the molokheiya leaves in LDLR (Low density lipoprotein receptor) /- mice fed high fat diet. The effect was associated with a reduction in oxidative stress and enhancement of B-oxidation in the liver. The results suggest that the consumption of molokheiya leaves may be beneficial for preventing diet induced obesity.

2.2.4.1.4.1.7 Blood chemistry effects Oyedeje and Bolarinwa [**33**] study on the effect of aqueous extract of *C. olitorious* (AECO) at doses of 250mg/kg, 500mg/kg, 750mg/kg BW on haematological and plasma biochemical parameters in male albino rats. 30 days with all doses of AECO caused insignificant changes in PCV, RBC, MCV, MCHC, MCH, platelet, neutrophil, lymphocyte, esinophil and monocyte values relative their respective controls, but caused significantdecrease in TWBC count relative to the control. Treatment 750mg/kg BW of AECO cause significant increase in total protein level relative to the control but others caused significant reduction globulin levels. It showed both harmful and beneficial potentialities on the blood chemistry. The effect on human blood chemistry is unknown. It may be recommended in its consumption by people with blood disorder. An increase in body weight (including weight of liver) was noticed in test animals after feeding with a protein-enriched diet from *C. olitorius* seeds. AST, ALT and total lipid of liver increased significantly whereas AST and ALT of serum decreased. A cholesterol free diet containing dried green leaves powder of*C. olitorius* lowered hepatic cholesterol concentration andincreased neutral fecal bile acid and neutral sterols excretion in rats [**34**]. Effects of multiple weekly dose of methanol extract of *C. olitorius* (15, 20, 25 mg/kg.i.p.) on liver and kidney functions and haematological parameters in mice were studied. No significant alteration of RBC count and hemoglobin content was observed in all dose levels of the treatment whereas significant increase in clotting time was seen in moderate and high doses. The extract caused significant increase in WBC count only at a high dose level of treatment. SGOT, SGPT, NPN and plasma cholesterol levels increased significantly at medium and high doselevels. Serum alkaline phosphatase and total bilirubin levels were also increased by both moderate and high dose levels of treatment. Low doses of the extract did not exhibit any significant change of creatinine and serum protein levels, but the high dose level significantly increased creatinine level [**35**].

2.2.4.1.4.1.8 Gastroprotective effect Rami and his colleagues [**36**] studied on the gastroprotective effect of ethanolic leaf extract of *C. olitorius* against ethanol induced gastric ulcers in adult Sprague dawley rats in animal model. The results show a gastroprotective property comparable to reference control drug omeprazole. Mezui and his colleagues [**37**] were find out the possible antiulcer activity of the extract of *C. olitorius* leaf in rats attributed to its ability to reduce acid secretion, to enhance mucosal defense and in vivo antioxidant status.

2.2.4.1.4.1.9 Antimicrobial and Synergism-Antagonism with antibiotics Semra and his colleagues [**38**] investigated the petroleum ether extract of *C. olitorius* a good activity against *E. coli, Staphylococcus aureus* and *Yersinia enterocolitica*. On the other hand ethyl acetate and water extract a good activity against *Geotrichum condidum* and *Botrytis cinerea*. Rume [**9**] experiment of the antibacterial and antifungal activity of the leaves methanolic extract of *C. capsularis*. It acts as against gram positive bacteria (*Bacillus subtilis, Staphylococcus aureus,* Beta-hemolytic *streptococcus, Bacillus cereus* and*Streptococcus pyrpgen*), gram negative bacteria (*Shigella boydii, Salmonella typhi, E. coli, Klebsiella* spp. and*Vibrio mimicus*), yeast and fungi (*Candida albicans, Saccharomyces cerevisiae* and *Bacillus megaterium*). Ashidi and his colleagues [**39**] described the in vitro interaction of an ethanol extract of leaf of*C. olitorius* with five antibiotics on methicillin sensitive and methicillin resistant *Staphyloccus aureus*. The extract synergized the antibacterial potential of ciprofloxacin and ampicillin/cloxacillin mixture and antagonized gentamycin, streptomycin and erythromycin on *S. aureus*.

2.2.4.1.4.1.10 Anticonvulsive and Arsenic induced toxicity / Protective effect properties Malaya and his colleagues [**40**] methanolic extract of both Cuscuta reflexa stem and *C. olitorius* seed showed marked protection against convulsion induced by chemo- convulsive agents in mice. Results suggest significant increases in catecholamines and GABA system, both considered to have significant roles with respect to CNS depressant and anticonvulsive properties of the processed extract. Das and his colleagues [**41**] find out the treatment with aqueous extract of *C. olitorius* leaves against sodium arsenite induced cardiotoxicity in experimental rats. The leaf has significant protecting effect against arsenic induced myocardial injury.

2.2.4.1.4.1.11 Atopic Dermatitis and Wound healing activities Satoshi and his colleagues [**42**] investigated the skin hydration effect of *C. olitorius* leaf extract excluding high molecular weight compounds on atopic dermatitis (allergic diseases like eczema) in mice. Results suggest that the leaf extract has therapeutic potential for atopic dermatitis due to its suppression of the plasma immunoglobulin E (IgE) level and degranulation of mast cells. Barku and and his colleagues [**10**] demented that *C. olitorius* displayed significant free radical scavenging activity and ferric reducing antioxidant power (FRAP).It also showed by excision model which proved that the traditional uses of this plan to treat wound has therefore been confirmed.

2.2.4.1.4.1.12 Antidiabetic and Bio- adsorbent properties Maxwell and his colleagues [**43**] studied the ethanolic seed extract of *C. olitorius* in animal model. They found that it has great potentials as an antidiabetic remedy due to the ability of the extract to lower blood glucose levels in normal rats, diabetic rats and also suppress postprandial rise in blood glucose level. Saliu and his colleagues [**44**] carried out of the supplementation of*C. olitorius* dried leaves as diet in type-2 diabetic rats. They showed that it has antihyperglycemic, antihyperlipidemic and antiperoxidative effects. It also exhibited modulatory effects on purinergic enzymes involved in the prevention of platelet abnormality and consequent vascular complications in diabetic state. Omeje and his colleagues [**45**] investigated the effect of *C. olitorius* extract on platelet aggregation. The results of the aggregation test showed that *C. olitorius* has the ability to aggregate blood plates. Sanjida and his colleagues [**46**] found that the methanol extract of*C. olitorius* aerial parts led to dose–dependent reduction blood glucose level in mice. Kingsley and his colleagues [**47**] investigated the effect of ethanolic extract of *C. olitorius* leaves to ascertain its anti-hyperglycemic property. Diabetes was induced intraperitorially, by injecting streptozotocin with a single of 50mg/kg body weight. After oral administration of extract for thirty day, the plant extract has high safety index as no toxicity or mortality was record at after the oral administration 5000 mg/kg body weight. The extract significantly decreased the activities of catalase, superoxide dismutase and liver function enzymes assayed alanine aminotransferase (ALT), alkaline phosphatase (ALP), and aspartate aminotransferase (AST) when compared to the untreated groups. The results indicate that the extract possesses anti- hyperglycemic effect. So, it could be used in the management diabetes mellitus. Subasri and his colleagues [**48**] evaluated the potential use of *C. olitorius* nano carbon as adsorbent for removal of Rhodamine–B dye (dye contaminated waste water passed out from industries like textile). The results suggest that the activated *C. olitorius* nano carbon (ACONC) may be utilized as a low cost adsorbent for Rhodamine–B dye removal from aqueous solution.

2.2.4.1.4.1.13 Anti-nutrients and Insecticidal effect Musa and Ogbadoyi [**49**] investigate on the anti-nutrient and toxic substance of *C. olitorius.* The concentration of cyanide in the fresh sample is lower (128mg/Kg) than maximum permissible level of 200mg/Kg fresh weight vegetables. This result indicates that the cyanide concentration in the fresh and pressed samples may not be high enough to induce toxicity in man. The concentrations of nitrate (3107 mg/Kg), soluble (3.49g/Kg) and total oxalates (5.85g/Kg) in fresh are higher than the permissible levels. The negative health problems associated with high intake of these photo-toxins can be averted though cooking and sun drying. With moderate cooking being the preferred method, since it reduce the plant toxins to tolerable level without compromising the nutritional potentials of this plant. Elangovan and his colleagues **[50]** observed among the experimental larval groups treated with leaves methanol extract of C. capsularis. They found that the minimum lethal concentration (LC_{50}) with 176.19ppm and 182.06ppm against *Anophele stephensi* (malarial vector) and *Aedes aegypti* (dengue vector). Other hand, with regard to the ovicidal activity of acetone, ethyl acetate and methanol extract of *C. capsularis* against the eggs of *An. Stephensi* and *Ae. aegypti*. The eggs exposed to ethylacetate and methanol extract were shown more susceptibility. The result showed that no hatchability 300-450ppm concentration on *An. Stephensi* and 375-450pp concentrations on *Ae. aegypti* respectively. So, it can be used as eco-friendly mosquito control. On the other hand, Roy **[51]** experiment of the role of jute leaf phytochemicals on feeding, growth and reproduction of *Diacrisia casignetum* Kollar (Lepidoptera: Arctiidae), it appeared that the larval and post larval developmental duration was shorter on mature jute leaf fed insects whereas adult longevity was higher in relative to young and senescent leaf fed insects. Fecundity of*D. casignetum* was also highest on mature leaves followed by young and senescent leaves. The growth and development of *D. casignetum* were related to the nutrient content relative to the secondary metabolites of these three types of jute leaves. Higher levels of nutritional factors (total carbohydrates, proteins, lipids, nitrogen and amino acids including water content) and lower levels of anti-nutritional factors (secondary metabolites) in mature jute leaves have influenced lower developmental time along with higher growth rate, fecundity and accumulated survivability of*D. casignetum* than the young and senescent leaves.

2.2.4.1.5 Conclusion

Jute is a well known bast fibre plant with less known medicinal scientific data. Each and every part of it is useful having medicinal activities. The folkoric medicinal value imparts tremendous value of this herb. On the hand, it may be to obtain key chemical compounds for making new drug molecules. The present information helps in bridging the gap between modern scientific studies and available traditional medical reports on this plant. As chemical constituents of *Corchorus* spp. vary depending on the origin and plant parts, a protocol needs to be standardized to isolate and obtain pure compounds. Though various chemical compounds have been isolated, purified and characterized, many compounds are yet to be studied in detail. Till now, limited efforts are made in the pharmacokinetics investigations related to the mechanism of action of the individual isolated compounds under in vivo conditions. Also, research should be emphasized on the toxicity and safety aspects of *Corchorus* spp. compounds using animal models. Further, therapeutic prospects of many new chemical compounds from *Corchorus* spp under in vitro and in vivo conditions should be explored in detail. In future, its aerial parts are used in multiple use as edible food products such as functional food.

2.2.4.1.6Reference

[**1**] Islam MM. 2013. Biochemistry, Medicinal and Food values of Jute (Corchorus capsularis L. and C. olitorius L.): A Review. I. J. of Enhanced Research in Science Tech. Eng. 2 (11): 35 -44.

[**2**] Mosihuzzaman M, Theander O and Amanb P. 1982. Analysis of Carbohydrates in the Jute Plant (Corchorus capsularis) . J. Sci. Food Agric. 33: 1207-12.

[**3**] Mosihuzzaman M, Fazlul HM and Chowdhury TA. 1988. Phenolic Acids in Fresh and Retted Jute Plants (Corchorus capsularis and C. olitorius L.). J. Sci. Food Agric. 42: 141 –47.

[**4**] Sarker SD and Nahar L. 2007. Chemistry for Pharmacy Students General Organic and Natural Product Chemistry. J Natural Product Chemistry. 1: 283 – 59.

[**5**] Bazzano LA, Ogden LG, Loria CM, Vupputuri S, Myers L. and Whelton PK. 2002. Fruit and vegetable intake and risk of cardiovascular disease in US adults: the first national health and nutrition examination survey epidemiologic follow up study. American J. Clinical Nutrition 76(1): 93-99.

[**6**] Costa M S, Duarte ARC, Cardoso MM and Duarte CMM. 2007. Supercritical antisolvent precipitation of PHBV microparticals. Inter. J. Pharmaceutics. 328 (1): 72 – 77.

[**7**] Ali MM, Nasreen A, Akhter S, Alam MA and Hossain A. 2013. Survey of indigenous knowledge of jute leaf in Bangladesh. Int. J. Sustain. Agri. Tech. 9(11): 1-4.

[8] Islam MM and Rahman M. 2008. In: Hand book on agricultural technologies of jute, kenaf and mesta crops. Bangladesh Jute Research Institute, Manik Mia Avenue, Dhaka-1207, Bangladesh. 92

[9] Rume JM. 2010. Phytochemical, antimicrobial and biological investigations of methanolic extract of leaves of Corchorus capsularis. Thesis for bachelor degree of pharmacy, East West University.

[10] Barku VYA, Boye A and Quansah N. 2013.Antioxidant and wound healing on the extrats of Corchorus olitorius leaf. World Essays J. 1(3): 67-73.

[11] Toshio F, Rong W, Katsuichiro O, Feroj AFMH, Idrish MA, Akira K and Hiroshi F. 2002. Antitumor promoters in leaves of jute (Corchorus capsularis and C. olitorius). Food Sci. Technol. Res., 8(3): 239-43.

[12] Hasan CM, Islam A, Ahmed M, Ahmed M and Waterman PG. 1984. Capsugenin, a dammarane triterpene from Corchorus capsularis. Phytochemistry. 23(11): 2583- 87.

[13] Ramadevi D and Ganapaty S. 2013. Phytochemical examination of Corchorus capsularis roots. Int. J. Pharmacognosy and Phytochemical Research. 5 (3): 173-76.

[14] Tabassum T. 2009. Extraction, identification and estimation of caffeine and catechin from Corchorus capsularis leaves extract. Thesis of a bachelor degree in pharmacy, East West University.

[15] Ali MM, Nasreen A, Sarker SC and Assaduzzaman M. 2014. Biochemical, physical and sensory evaluation of Jute leaves (Corchorus Spp) as affected by storage periods after direct sundry and hot water treatments. Bangladesh J. Jute Fib. Res. 31 (1-2): 75-80.

[16] Ali MM, Hossen MZ, Akhter S and Tahmina. 2015. Study on storage properties of jute leaves by blanching method. Int. J. Sustain. Agri. Tech. 11(12): 12-16.

[17] Rao EV, Rao DV, Pavanaram SK, Von Euw J and Reichstein T. 1971. Structure of Corchoroside β-Glycosides and aglycones. Helv Chim ACta. 54(7): 1960-68.

[18] Rao MR Liu TP and Meng ZJ. 1979. Comparative studies on the cardiotonic actionof corchoroside A,ouabain and strophanthim K on the heart–lung preparation of guinea pig and cat. Yao Xue Xue Bao. 14(5): 257– 66.

[**19**] Del Rio JC, Marques G, Rencoret J, Jimenez–Barbero J, Martinez AT and Gutierrez A. 2009. Chemical composition of lipophilic extractives from jute (Corchorus capsularis) fibers used for manufacturing of high quality paper pulps. Industrial Crops and Products 30(2): 241–49.

[**20**] Manzoor–i-Khuda M and Gerhad H. 1979. Chemical constituents of Corchorus capsularis and C. olitorius (jute plant part IV, isolation of corosolic acid, ursolic acid and oxo–corosin and correlation of corosin with tormentic acid. Z. Naturforsch. 29: 1320-25.

[**21**] Ali MM and Nasreen A. 2016. Sensory evaluation of dried jute leaf as vegetable and herbal tea. Int. J. Sustain. Agri. Tech. 12(1): 23-26.

[**22**] Zakaria ZA, Sulaiman MR, Gopalan HK, Abdul Ghoni ZD, Raden MNRN, MatJAM, Abdullah C. 2007. Antinociceptive and antiinflamatory properties of Corchorus capsularis L. leaves chloroform extract in experimental animal model. J. Pharmaceutical Society of Japan. 127(2): 359 –65.

[**23**] Kimiko O, Kouji O, Uki Y, Isao Y and Akira M. 1995. Characterization of an acidic polysaccharide isolation from the leaves of Corchorus olitorius (Moroheiya). Biosci. Biotech. Biochem., 59(3): 378 -81.

[**24**] Özlem Dİ, Erkan Y, Feride IS and Mehmet H. 2013. Corchorus olitorius (jute) extract induced cytotoxicity and genotoxicity on human multiple myeloma cells (ARH-77). Pharm Biol, 51(6): 766–70.

[**25**] Mohamed IMI, Rialet P, Skena HAA, Anna MVWCCB, John PG, Masaad A and Abdel W. 2015. Aqueous extract of Corchorus olitorius decreases cytotoxicity of aflatoxin B1 and fumonisin B1 in H4 IIE –luc cells. Hepatoma Research 1(2): 75-85.

[**26**] Azuma K, Nakayama M, Koshioka M, Ippoushi K, Yamaquchi Y, Kohata K, Yamauchi Y, Ito H. and Hiqashio. 1999. Phenolic antioxidants from the leaves of C. olitorious L. J. Agric. Food Chem. 47(10): 3963-66.

[**27**] Zakaria ZA, Sulaiman MR, Gopalan HK, Abdul GZD, Raden MNRN, Mat JAM, Abdullah C. 2007. Antiiflammatory and Antipyretic activity of C. olitorius in rats. J. Pharmaco. Toxicol. 1: 134-40.

[**28**] Zainul AZ, Sulaiman MR, Gopalan HK, Abdul Ghani ZDF, Raden MNRNS, Mat JAM and Abdullah FC. 2007. Antinocicceptive and anti-inflammatory properties of Corchorus capsularies leaves chloroform extract in experimental animal models. Yakugaku Zasshi. 127(2): 359 –65.

[**29**] Frerejacque M and Durgeat M. 1954. Digitalis like poisons of jute seed. Compt Rend. 238: 507–509.

[**30**] Karrer P and Banergee P. 1949. Corchortoxin a cardiac agent from jute seeds. Helv Chim Acta. 32 : 2385-92.

[**31**] Negm S, El- Shabrawy O, Arbid M and Radwan AS. 1980. Toxicologyical study of the different organs of Corchorus olitorius L. plant with special reference to their cardiac glycosides content. Zeitsc Ernaehrungsw. 19(1): 28-32.

[**32**] Wang L, Masyiuki Y, Takuya K, Xufeng S, Yukikazu Y and Kuninori S. 2011. Anti-obesity effect of polyphenolic compounds from molokheiya (C. olitorious) leaves in LDL receptor deficient mice. European J. Nutrition. 50(2): 127-33.

[**33**] Oyedeje KO and AF Bolarinwa AF. 2013. Effect of C. olitorius extract on haematological and plasma biochemical parameters in male albino rats. JOSR J. Dental and Medical science. 3(5): 68-71.

[**34**] Innami S, Tabata K, Shimizu J, Kusunoki K, Ishid MM, Wada M, Sugiyama N and Kondo M, Drie. 1998. Green leaf powders of Jew's mellow (Corchorus olitorius), persimmon (Diosphyros kaki) and sweet potato (Ipomoea batatas Poir) lower hepatic cholesterol concentration and increase fecal bile acid excretion in rats fed a cholesterol free. Plant Foods Hum Nutr. 52(1): 55–65.

[**35**] Mazumder UK, Gupta M, Pal DK and Bhattacharya S. 2003. Chemical and toxicological evaluation of methanol extract of Cuscuta reflexa Roxb. stem and Corchorus olitorius Linn seed on haematological parameters and hepatorenal functions in mice. Acta Pol Pharm. 60: 317-23.

[**36**] Rami AB, Fouad AB, Mahmood AA, Mazen MJ, Al- obaidi, Maryam H, Pouya H, Mustafa F, Shahram G and Samaneh T. 2013. Gastoprotective effects of C. olitorius leaf extract against ethanol induced gastric mucosal hemorrhagic lesions in rats. J. Gastroenterol. Hepatol. 28(8): 1321-29.

[37] Mezui C, Amang AP, Nkenfou C, Sando Z, Betou D, Moulioum H, Tan PV. 2013. Anti-ulcer and antioxidant activity of the leaf aqueous extract of Corchorus olitorius (Tiliaceae) in rats. J. homepage: www.onlineijp.com

[**38**] Semra I, Filiz S and Ferdag C. 2007. Antibacterial and antifungal activity of C. olitorius extracts. Int. J. Natural and Eng. Sci. 1 (3): 59-61.

[**39**] Ashidi JS, Efuntoye MO, Odunbaku OA and Biliaminu SA. 2012. The in vitro effect of Corchorus olitorius on the antibacterial activities of five antibiotics. Global Research J. Microbiology. 2(2): 113 –17.

[**40**] Malaya G, Upal KM, Dilipkumar P, Shailaditya B and Sumit C. 2003. Studies on Brain Biogenic Amines in Methanol Extract of Cuscuta refexa and Corchorus olitorius Seed Treated Mice. Acta PoliniacPhamaceutica-Drug Research. 60 (3): 207-10.

[**41**] Das AK, Sahu R, Dua TK, Baq S. Gangopadhyay M. Sinha MK and Dewanjee S. 2010. Arsenic Induced Myocardial Injury: Protective role of C. olitorius leaves. Food Chem Toxicol. 48(5): 1210 -7.

[**42**] Satoshi Y, Keiichi H, Takahiko F, Hiroya K, Nobuyuki K, Shu S, Makoto I and Kazuya O. 2014. Skin Hydration Effects of C. olitorius leaf extract in mouse model of atopic dermatitis. J. Cosmetics, Dermatological Sci and Applications . 4: 1-6.

[**43**] Maxwell OE, Emmanuel UE, Shaibu OB and Sanusi WH. 2013. Anidiabetic activity of ethanolic seed extract of C. olitorius. Int. J. Sci : Basic and Applied Research (IJSBAR). 12(1): 8-21.

[**44**] Saliu JA, Oboh G, Schetingeer MR, Stefanello N, Rocha JBT. 2015. Antidiabetic Potentials of Jute Leaf (C. olitorius) on type-2 diabetic rats. J. Emerging Trends in Engineering and Applied Sci. 6 (7): 223-30.

[**45**] Omeje KO, Odiba AS, Ejembi DO, Chukwuka RS. 2014. Platelet Aggregation Effect of C. olitotorus extract as an index of managing blood clotting Disorders. J. Dental and Medical Sci. 13 (10) 58-60.

[**46**] Sanjida P, Moytry M, Sanjaana R, Anul KD, Sanjida H and Mohammed R .2015. Preliminary phytochemical screening, antihyperglycemic, analgesic and toxicity studies on methanolic extract of aerial parts of C. olitorius. J. Applied Pharm. Sci. 5(9): 68 –71.

[**47**] Kingsley OO, Tobechukwu CE, Henry CO and Benjamin OE. 2013. Effect of ethanol extract of Corchorus olitorus leaf on glucose level and antioxidant enzymes of Streptozotocin induced hyperglycemic rat. Nigerian Society for Experimental Biology (http://nisebpublications. Org).

[**48**] Subasri S, Arivoli S, Marimutha V and Mani N. 2015. Equilibrium, Kinetic and Thermodynamic Study on Rhodamine–B Removal from Aqueous Solution Using Activated C. olitorius leaves. Inter. J. Plant, Animal and Environmental Sci. 5(1): 208–18.

[**49**] Musa A and Ogbadoyi EO. 2012. Effect of Cooking and Sun Drying On Micronutrients, Antinutrients and Toxic Substances in Corchorus Olitorius (Jute Mallow). J Nutr Food Sci 2:14.

[**50**] Elangovan A, Dhanasekaran S, Anandan A, Krishnappa K, Gokulakrishnan J and Elumalai K.2012. Mosquitocidal activities of Corchorus capsularis L (Malvaceae) against a common mal aria vector, Anopheles stephensi (Liston) and a dengue vector Aedes aegypti (L) (Diptera: Culicidae). I. J. Recent Scientific Research. 3(6): 564-8.

[**51**] Roy N. 2014. Role of Chorchorus capsularis phytochemicals on the feeding dynamics of Diacrisia casignetum Kollar (Lepidoptera: Arctiidae). J. Entomology and Zoology Studies. 2 (4): 227-36.

2.2.4.2Jute leaf- A powerful biological tool and food supplement

[Zakaria et al 2019, Md. Mahabub et al 2019, Aleya et al 2021, Zakaria and Shuranjan 2022, Taslima et al 2022, Aleya et al 2022]

2.2.4.2.1 Introduction

The increasing proliferation of diseases, with the associated drug resistance, has underscored the need for new drug development. Plant-derived natural products possess diverse pharmacological activities and consequently are an attractive resource for the development of advanced chemotherapeutics to treat a wide range of disease conditions, including microbial infections. Jute leaves rich with active diterpenoids for the potential discovery of new drugs and boost up the immune system. It is commonly used for its anti-inflammatory, antineoplastic, antipyretic, anti-hypertensive, antidiabetic, cytotoxic and antipyretic effects and immune-boosting properties. Therefore, in the search for economically feasible with better efficacy and low toxicity, plant sources are highly prioritized. *Corchorus* spp. (jute) is a native plant of tropical Africa and Asia, and has since spread to Australia, South America and some parts of Europe. Jute leaves not only make great ingredients but they have also been found to have numerous medicinal benefits.Jute leaf is rich in secondary metabolites and is used in folk medicine in jute-producing communities [**1**].

2.2.4.2.2. Constituents and physical properties of jute leaf More than 80 compounds, including glycosides, triterpenes, ionones, phenolics, phytosterols, organic acids, lignins, alkaloids have been isolated and identified from jute plant [**2, 3**]. Leaves of jute plants are 6-10cm long and 3.5-5cm broad. Leaves alternate, simple; stipules narrowly triangular with long point; petiole 1–7cm long; blade narrowly ovate, ovate or elliptical, 4–15cm×2–5cm, cuneate or obtuse and with setaceous appendages up to 2.5 cm long at base, acuminate to acute at apex, margin serrate or crenate, almost glabrous, usually shiny dark green, 3–7 veined from the base. Inflorescence a 1–4 flowered axillary fascicle, bracteate. Flowers bisexual, regular, usually 5 merous, shortly stalked; sepals free, narrowly obovate, 5–7mm long; petals free, obovate, 5–7mm long, yellow, caducous; stamens numerous; ovary superior, usually 5-celled, style short from a cylindrical capsule up to 7–10cm long, ribbed, with a short beak, usually dehiscing by 5 valves, many-seeded. Seeds angular, 1–3mm long, dark grey. Seedling with epigeal germination; hypocotyl 1–2cm long;

cotyledonsfoliaceous, broadly elliptical to circular, 3–8 mm long [**4**]. Leaves are normally light green colored and are bit bitter in taste. When collected young, jute leaves are tasty and tender; older leaves tend to be more woody and fibrous, making them less ideal for consumption. In addition to adding a distinctive flavor to food, jute leaves also have nutritional value, and they act as thickeners in soups, stews, and sauces.Apart from their slightly bitter taste jute leaves is a good source of nutrients, vitamins and minerals (**Table 2.2.4.2.1 and 2.2.4.2.2**) [**1, 2, 5, 6**]. Apart from food use, it is grown as a commercial crop for jute production in India, Bangladesh and China [**7**].

Table 2.2.4.2.1 Properties of jute leave (*C. capsularis*). [Zakaria and Shuranjan 2022]

Constituents	**Per 100 g of the leaves**
Calories	43-58
Moisture (water)	80.4-84.1 g
Protein	4.5-5.6 g
Fat	0.3 g
Total carbohydrate	7.6-12.4 g
Fibre	1.7-2.0 g
Ash	2.4 g
Ca	266-366 mg
P	97-122 mg
Fe	7.2-7.7 mg
Na	12 mg
K	444 mg
β-carotene equivalent	6.41-7.85 mg
Thiamine (Vitamin B_1)	0.13-0.15 mg
Riboflavin (Vitamin B_2)	0.26-0.53 mg
Niacin	1.1-1.2 mg
Ascorbic acid (Vitamin C)	53-80 mg

Table 2.2.4.2.2 Phytochemical composition of *C. olitorius*[Zakaria and Shuranjan 2022]

Phytochemicals		References
Strophanthidin glycosides		
1	Erysimoside(strophanthidin 3-O-β-D-glucopyranosyl (1→4)-O-β-D-digitoxopyranoside),	[8, 9, 10]
	Olitoriside (strophanthidin 3-O-β-D-glucopyranosyl (1→4)-O-β-D-boivinopyranoside),	
2	Corchoroside A (strophanthidin 3-O- β-D-boivinopyranoside) and helveticoside (strophanthidin 3-O- β-D—digitoxopyranoside),	
3	Corchorusoside (A, B, C, D, and E)	
Digitoxigenin glycosides		
1	Glucoevatromonoside (digitoxigenin-3-O-β-D-glucopyranosyl-(1→4)-O-β-D-digitoxopyranoside),	[8, 9]
2	Coroloside(digitoxigenin-3-O-β-D-glucopyranosyl(1→4)-O-β-D-boivinopyranoside),	
3	Deglucocoroloside(digitoxigenin-3-O-β-D-boivinopyranoside),	
4	Evatromonoside(digitoxigenin-3-O-β-D-digitoxopyranoside),	
	Digitoxigenin 3-O-β-D-glucopyranosyl-(1→ 6)-O-β-D-glucopyranosyl-(1→4)-O-β-D-digitoxopyranoside	
Cardiac glycosides		
1	Cannogenol-3-O-β-D-glucopyranosyl-(1→4)-O-β-D-boivinopyranoside,	
2	Periplogenin-3-O-β-D-glucopyranosyl-(1→4)-O-β-D-digitoxopyranoside	[11]
Phenolics		
1	5-caffeoylquinic acid (Chlorogenic acid)	[12, 13, 14]
2	(2) 3,5-dicaffeoylquinic acid (isochlorogenic acid)	
Flavonoids		
1	Astragalin (kaempferol 3-O-β-D glucopyranoside)	[13, 15, 16]
2	Tolifolin (kaempferol 3-O-β-D-galactopyranoside)	
3	Jugalanin (kaempferol 3-O-α-L-arabinopyranoside)	
4	Isoquercetin (quercetin 3-O-β-D-glucopyranoside)	
Coumarins		
1	Cichoriine	[12, 17]
2	Scopolin	
3	Saponins	
4	Tannins	

Triterpenes		
1	Ursolic acid, Corosolic acid, Oxocorocin	[18, 19, 20]
2	β-Sitosterol	[21]
3	(6S,9R)-roseoside	[6, 10]
4	3,4-di-O-caffeoylquinic acid	[22, 23]
5	3,5-dicaffeoylquinic acid	[10, 14]
6	4-O-caffeoylquinic acid	[22, 23]
7	Alkaloids	[6, 26]
8	Apigenin-7-O-glucoside	[22, 23]
9	Apigenin	[22, 23]
10	Astragalin	[10]
11	Betulabuside A	[10]
12	Caffeic acid	[22, 23]
13	Campesterol	[6, 24]
14	Carvacrol methyl ether	[24]
15	Cedran-5-one	[6, 24]
16	Chlorogenic acid	[10, 14]
17	Cichoriine	[10]
18	Cirsilineol	[22, 23]
19	Cirsiliol	[22, 23]
20	Cis-β-dihydroterpineol	[24]
21	Corchoiononside A , B, C	[6, 10]
22	Corosolic acid	[20]
23	Eicosane	[6, 24]
24	Ethyl salicylate	[24]

25	Ferulic acid	[22, 23]
26	Gingerol	[22, 23]
27	Heptadecane	[6, 24]
28	Heptadecanoic acid	[24]
29	Hexadecane	[24]
30	Isobutyl salicylate	[24]
31	Isochlorogenic acid	[10, 13, 25]
32	Isoquercetin	[10]
33	Jugalanin (kaempferol 3-O-α-L-arabinopyranoside)	[25]
34	Kaempferol	[22, 23]
35	Linoleic acid	[24]
36	l-menthone	[24]
37	Luteolin	[22, 23]
38	Methyl tiglate	[24]
39	Myricetin	[22, 23]
40	Naringenin	[22, 23]
41	Naringin	[22, 23]
42	Nonadecane	[24]
43	Octadecane	[24]
44	Oleanolic acid	[6, 25]
45	Oxocorocin	[20]
46	Palmitic acid	[24]
47	P-coumaric acid	[22, 23]
48	Piperonal	[24]
49	Protocatchuic acid	[22, 23]
50	Quercetin	[22, 23]
51	Quercetin-3-(6-malonylgalactoside)	[14, 25]
52	Quercetin-3-galactoside	[14, 25]
53	Quinic acid	[22, 23]
54	Rosmarinic acid	[22, 23]
55	Rutin	[22, 23]
56	Saponins	[17, 26, 27]
57	Scopolin	[10]
58	Stearic acid	[24]
59	Stigmasterol	[24]
60	Tannins	[17, 23, 27]
61	Tetradecane	[24]
62	Tolifolin (kaempferol 3-O-β-D-galactopyranoside)	[25]

63	Trans-cis-Farnesol	[24]
64	Trans-Phytol	[6, 24]
65	Trans-β-dihydroterpineol	[24]
66	Vanillic acid	[22, 23]
67	β-amyrin	[24]
68	β-sitosterol	[21, 24]
69	1,2-benzenedicarboxylic acid	[28]
70	1,2-benzenedicarboxylic acid, dibutyl ester	[28]
71	Aldehyde	[28]
72	Benzaldehyde	[28]
73	Camphene	[28]
74	Cyclohexane	[28]
75	Fraxinellone	[28]
76	Geranylisobutyrate	[28]
77	Geranyl propionate	[28]
78	Germacrene D	[6, 28]
79	Heneicosane	[28]
80	Hexadecanoic acid	[28]
81	Hexanoic acid, methyl ester	[28]
82	Hexanone	[6, 28]
83	Hexenyl benzoic acid	[28]
84	Hydrocarbons	[28]
85	Isoamyl butyrate	[28]
86	Limonene	[28]
87	Methyl tiglate	[28]
88	Nerolidol	[28]
89	Nonadecane	[6, 28]
90	Octadecanoic acid	[28]
91	Phenyl ethyl tiglate	[28]
92	Sabinene	[28]
93	Terpenes	[28]
94	α-phellandrene	[28]
95	α-pinene	[6, 28]
100	Phytol	[29]
99	Monogalactosyldiacylglycerol	[6, 29]
98	β-myrcene	[28]
97	β-cedrene	[28]
96	α-terpinene	[28]

Table 2.2.4.2.3a Vitamin C and beta carotene content of jute leaves (mg/100 g dry weight basis). [Zakaria and Shuranjan 2022]

Component	Treatment			
	Fresh	Shade dried	Solar dried	Cooked
Vitamin C	135.60±8.75	16.82±0.45	11.55±0.24	24.45±0.95
Beta carotene	7.82±0.16	4.65±0.32	4.42±0.28	3.81±0.42

Table 2.2.4.2.3b Qualitative presentation of phytochemicals in different condition of the jute leaves. [Zakaria and Shuranjan 2022]

Treatment	Phytochemical compounds							
	Flav	Alka	Sap	Tan	Phe	Ant	Ster	Ter
Fresh	+	+	+	+	+	+	+	-
Shade	+	+	+	+	+	±	±	-
Solar	+	+	+	+	+	-	-	-
Cooked	+	+	+	+	+	-	-	-
+ Present, – Absent, ± Doubtful								
Alka-alkaloid, Flav-flavonoid, Sap-saponin, Tan- tannin, Phe- phenol, Anth- anthraquinone, Ster-steroids, Terp-terpenoids.								

Deferent experiments have revealed that plant parts, such as leaves, stems, roots, barks, and seeds, of *C. capsularis* and *C. olitorius* contain polysaccharides, flavonoids, phenolics, cardiac glycosides, sterols, fatty acids, triterpenoids, and ionones [**30**]. Ahmed [**6**] reported that *C. olitorius*leaves are rich sources of Vitamin-A (β-carotene), C, E, B_1, B_2, folic acid and minerals such as iron and calcium in addition to common macromolecules [**2, 31, 32, 33**] (**Table 2.2.4.2.3a, 2.2.4.2.3b and 2.2.4.2.3c**). Folic acid (folate, vitamin-B) substantially higher (0.03mg/L) than that of other folacin rich vegetables and iron 72µg/g. Jute leaves yielded a lot of mucilaginous polysaccharide (acidic polysaccharide) and anthocyanin. Leaf extracts yielded alkaloids, terpenoids, tannins, flavonoids, glycosides [**34**]. It contains two functional compounds, phytol and monogalactosyldiacylglycerol that protect from free radicals [35]. Capsin and capsugenin isolated from the leaves of *C. capsularis* where capsin are responsible for the major bitter taste [36]. β-Sitosterol, scopoletin and fusidic acid also isolated from the leaves [**37**]. Caffeine and catechine isolated from *C. capsularis* leaves extract [**38**]. Oleanolic acid was isolated from the leaves of *C.*

olitorius of Egyptian origin. Four triterpenoid glycosides (chorchorusins A, B, C and D) isolated from aerial parts of *C. acutangulus*. Leaves of *C. olitorius* contain ionone glucosides which are corchoionosides A, B and C, an ionone glucoside (roseoside) and a monoterpeneglucosidebetulabuside A. Leaves of *C. olitorius* gave four higher fatty acids with a trienone system, corchorifatty acids A, B, C, D, an uncanoic acid, corchorifatty acid E and a trihydroxy fatty acid, corrchorifatty acid F [**39**]. Air dried powdered leaves of *C. capsularis* in chloroform in the ratio of 1:20 ($^{w}/_{v}$) demonstrated the presence of flavonoids, saponins, tannins, steroids and triterpenes [**31, 40**]. The leaves of *C. capsularis* gave a new dammaranetriterpine glycoside, capsin [**36, 41**]. Moreover, triterpineglucosidecapsugenin 30-O-glucopyranoside 36 was isolated from the mature leaves of the species [**42**].

Table 2.2.4.2.3c Vitamin and Proximate composition (%) of different parameters of *C. olitorius* leaf. [Zakaria and Shuranjan 2022]

Proximate composition (%)					
Moisture	Ash	Crude Fibre	Protein	Lipid	Carbohydrate
79.98 ± 0.22	0.64 ± 0.02	0.33 ± 0.01	6.21 ± 0.04	5.07 ± 0.02	6.25 ± 0.03
Vitamin composition (mg/100 g dry weight)					
Ascorbic acid	Niacin	Riboflavin	Thiamin		
316.80 ± 0.22	0.61 ± 0.02	0.06 ± 0.01	0.04 ± 0.01		

C. olitorius leaves contain macro and micronutrients as well as polyphenols [**14, 13, 16**]. Among carbohydrates, acidic polysaccharides are of particular interest because of their notable biological effects including antidiabetic and antioxidant. It is also a good source of a diverse category of phytochemicals including alkaloids, saponins, tannins, terpenes, flavonoids, and phenolics. Different extracts exhibit potent antidiabetic, antioxidant, antiinflammatory, anticancer, antimicrobial, hepatoprotective, cardioprotective, neuroprotective, analgesic, and wound healing effects [**6**]. The genus *Corchorus* is rich in bioactive molecules such as glycosides, polysaccharides, triterpenes, ionones, phenolics, sterols, and fatty acids [**43**]. These biomolecules, which are part of a large and varied group of chemical compounds, impart a range of prophylactic and therapeutic applications to the genus. Growing clinical evidence also supports the traditional uses of *Corchorus* as folk medicine. Given the immense genetic diversity, geographical adoptability and biological effects on human health, the genus *Corchorus* can serve as a potential herbal drug-source for enhancing community health standards and livelihood security of the rural population of Asia and Africa [**43**]. A study was carried out by Hassan et al [**44**] for the determination of lipid contents (fatty acids, phytosterols and hydrocarbon components) of seeds, roots, leaves and stems of *C. olitorius* as well as the seeds and vegetative part of *C. capsularis* which showed that cholesterol was the major compound. Research has also shown that these copper-rich leaves can help to reduce the levels of bad cholesterol (LDL) and increase beneficial cholesterol. Jute leaves have about 0.222mg of copper which can dramatically help to reduce chances of cardiovascular diseases as well as promote a healthy weight. For people suffering from sleeping disorders like sleep apnea and insomnia, jute leaves might be of help. Jute leaves are rich in magnesium which is the mineral that helps to treat and improve such conditions. Magnesium aids in the release of hormones that relax and soothe the nerves. By increasing the magnesium intake, you can improve your sleep quality for restful and uninterrupted sleeping patterns. The skin and cells require vitamin-A which aids in promoting wounds healing and regrowth of skin. Vitamin-A is essential in the growth of all epithelial cells both externally and internally and it also helps to prevent skin cancer. Jute leaves have an abundance of Vitamin-A. Eating them regularly can help prevent various skin conditions like poor complexion, acne, lines and wrinkles keeping your skin looking young and healthy. The Vitamin-A contained in jute leaves also helps to promote healthy growth of hair as well. The benefits of eating jute leaves extend to the teeth and gums. They contain calcium which helps in keeping the jaw bone strong and robust. A healthy jaw bone keeps your teeth tightly fit and makes it harder for bad bacteria to grow. To keep your teeth and gums from giving you any trouble, you should maintain a calcium-rich diet which should include daily consumption of jute leaves, especially for young children. Restless leg syndrome is a common condition usually caused by iron deficiency. Because jute leaves are rich in iron, eating them regularly can help to improve the condition. Research shows that the jute plants contain up to 2.73mg of iron which is more than a third of the daily recommended value. Eating jute leaves can also help to curb muscle spasms which are also considered to be caused by a lack of enough iron. Jute leaves are not only a delicacy in many countries. They are also a super food with numerous health and medical benefits that promote the general well-being of the body. If you're looking to try new and healthy foods, you should consider adding jute leaves to your diet. They are available fresh and dried and both varieties have the same benefits [**45**]. Singh et al [**46**] conducted a research to evaluate the suitability of *C. olitorius* leaf powder, as a dietary protein source for fish, Labeorohita fingerlings and found that C. olitorius leaf powder can be incorporated in the feed of rohu fingerlings up to 20% (significantly P<0.05 higher). Tareq et al [**47**] examined the nutritional components of jute leaf of *C. capsularis*, germplasm (Merha red, Merha green, Birol red), *C. olitorius* accession (Acc-3840), BJRI deshi pat shak-1 and BINA pat shak-1 (**Table- 2.2.4.2.4a and 2.2.4.12.2.4.2.4b**).

Table 2.2.4.2.4a Mineral composition of jute genotypes [47]

Genotype	Mineral composition				
	Ca%	K%	Na%	P%	Fe ($mg{\cdot}kg^{-1}$)
Merha red	2.36a[a]	1.61ab	0.125a	0.62ab	971c
Merha green	1.93ab	1.67ab	0.118a	0.588b	610d
Birol red	2.21a	1.97a	0.099a	0.613ab	1253b
Acc-3840	2.07ab	1.68ab	0.098a	0.655a	1208b
BJRI deshi pat shak-1	1.62b	1.75ab	0.111a	0.654a	993c
BINA pat shak-1	1.91ab	1.39b	0.091a	0.636ab	1344a
LSD	0.5304	0.3728	0.0575	0.0575	59.97
CV%	14.44	12.26	18.56	5.00	3.10

[a]Values in columns followed by the same letter are not significantly different, $p \leq 0.05$, LSD.

Table 2.2.4.2.4b Vitamin composition of jute genotypes.[Zakaria and Shuranjan 2022, Tareq et al 2019]

Genotype	Vitamin composition				
	Moisture (%)	Protein (%)	Ash (%)	β -carotene ($\mu g{\cdot}g^{-1}$)	Vitamin-C (mg/100 g)
Merha red	85.04a [a]	18.45c	7.770a	125.4a	74.53a
Merha green	82.12cd	22.54a	5.673c	127.5a	75.80a
Birol red	82.57bc	22.68a	6.837b	117.9b	59.83d
Acc-3840	83.06bc	20.72b	5.587c	103.7c	67.30b
BJRI deshi pat shak-1	84.08ab	22.66a	8.580a	115.8b	67.73b
BINA pat shak-1	80.63d	22.83a	5.690c	95.59d	63.85c
LSD	1.725	0.656	0.863	4.161	2.573
CV%	1.14	1.67	7.09	2.00	2.08

[a]Values in columns followed by the same letter are not significantly different, $p \leq 0.05$, LSD.

2.2.4.2.3. Food uses Jute leaves contains high amounts of mucilaginous polysaccharides which yield viscous soup when cooked and usually used as an accompaniment for main dishes [**48**]. In Middle Eastern countries, the leaves are cut into small pieces and boiled in water with salt and pepper to make soup. Jute leaves soup is very popular in the Middle East. In Mediterranean regions, young green leaves and shoots are used to add flavor and viscous texture to soups and stews. Seeds are used for flavoring. Tender leaves and shoots are also eaten raw as salad vegetable in Egypt and India [**49**]. Dried leaves are used in the preparation of herbal tea. The leaves are used to prepare a stew called "ewedu" in Nigeria, while in Philippines the leaves along with bamboo shoots are consumed as a leafy vegetable. In Nigeria, sticky sauce comparable to okra is prepared and eaten as an accompaniment for starchy dumplings made from cassava, yam or millet. Since, Jute leaf is an annual herb dried leaf powder is used to make this sauce during off season. Sauce is also prepared from powdered and dried immature fruits (bush okra). In East Africa, it is cooked with cowpeas, pumpkin, cocoyam leaves, sweet potato, milk, butter, and meat flavored with pepper and lemon [**20, 50**]. Recently, Jute leaves are also used for the development of Sushi wrap as a promising viable substitute for Nori [**51**].

2.2.4.2.4. Medicinal Benefits Medical science has made a remarkable advancement through the discovery of numerous drugs and medicines-antibiotics, hormones and vaccines-for the treatment of several diseases caused by pathogenic microbes, even though these pharmaceutically formulated medicines are not totally free from side effects. Nevertheless, more studies are still needed to be done to discover cure for many diseases that still remain incurable. This therefore underscores the need to involve herbal medicine in the fight against diseases, because it is cost effective with little or no side effects. To this end, different plants and herbs should be studied to harness their medicinal values. Studies have shown that plants are the major raw materials used in medical researches for the production of drugs and medicines. In fact, the use of plants in the management and treatment of diseases started with life. Human depends on plants for medicine; both fororthodox and herbal medicine. Through phytochemical screening of plants, it has been revealed that various plants have medicinal values and properties for the treatment of different human diseases. Jute leaves contain 0.0496mg of vitamin B_6 which is over 30% of the recommended daily value. Most eye diseases are a result of poor diet and nutrient deficiencies. Some of the nutrients that are linked with improving eye health and preventing eye diseases and disorders include Vitamin B_6 and folate among other vitamins. Regular consumption of jute leaves can help in slowing the onset of some eye diseases including age-related muscular degeneration. Patil and Jain [**27**] stated that successive solvent extracts of*C. olitorius* contains medicinally important bioactive compounds. Phytochemical analysis revealed the presence of phenols, flavonoids, tannins, saponins, alkaloids, fixed oil and fats whichconcluded that the crude extract of *C. olitorius* is a potential source of various activates and this justifies its use in folkloric medicine [**27**]. Vitamin-C which is abundant in jute leaves is essential in helping the immune system ward off colds and viruses. In addition to fighting colds, vitamin-C can also keep us from suffering further complications like lung infections and pneumonia. Jute leaves are rich in fibre too. Dietary fibre is one of the major elements that help in regulating the digestive process. By regularly eating jute leaves, you can avoid constipation, bulk up stool and speed up digestion. Having a fibre-rich diet also helpsto increase the uptake efficiency of nutrients and it reduces symptoms like cramping and gas [**45**]. Ali et al [**39**] reported that jute can be a potential medicinal product for the treatment of many diseases. Its leafy vegetable is popularly used in folk medicine for the treatment of fever, chronic cystitis, cold and tumours. There are many health benefits of jute leaf, such as-weight loss (due to the presence fatty acids, Omega 3, Vitamins-B, magnesium and zinc), prevents aging (due to antioxidant activity and also it prevents the loss of water through the skin), strengthens bone and teeth (due to the presence of calcium and phosphorus), stabilizes blood pressure (due to Omega 3), helps in hormone stabilization in women and improves brain health (due to Omega 3) [**52**]. One of the most celebrated medical benefits of jute leaves is their ability to battle inflammation in the body. Because they contain a variety of antioxidants, eating jute leaves can help to protect against damage caused by inflammation. It is also believed that jute leaves have anti-inflammatory properties. In a study conducted in 2015, it was found that the jute plant to have properties that can treat inflammation. By eating jute leaves regularly, one can prevent inflammation [**45**]. Jute leaves have a variety of properties that are critical in helping manage body weight. Their high fiber content is essential in improving digestion and keeping people feeling full for longer. On the other hand,*C. capsularis* leaves showed several pharmacological effects such as anticancer [**29**], anti-inflammatory, antinociceptive, antipyretic [**53**] and antimicrobial [**54**]. Nasreen et al [**55**] reported that jute leaves contain β-carotene but it varies among different varieties jute leaves and most of the varieties in *C. olitorius* showed higher β-carotene than the varieties of *C. capsularis*. This β-carotene is a carotenoid pigment that converts to vitamin-A in human body and plays crucial role in visual functions, reproductive performance and immune system. Health-flourishing effects of plant-derived secondary metabolites in human health, including antioxidative, anticarcinogenic, antibiotic, and pharmacological effects, are well documented [**56**]. *C. olitorius* leaves are used in the treatment of fever, tumors, pectoral pains, dysentery, aches, enteritis, cystitis, piles and dysuria [**57**]. *C. capsularis* leaves are also used in ayurvedic for ascites, piles, cystitis, dysuria, fever and gonorrhoea [**31**].

2.2.4.2.5. Exhibit potential for disease control Ahmed et al [**58**] stated that *Corchorus* spp. can be a potential medicinal product for the treatment of many diseases. In traditional medicinal practices, it is used to treat constipation, demulcent, dysentery, worm, carminative anthalmitic, intestinal antiseptic, ascites, pain, piles, tumors, dysuria, febrifuge, stomachic, cystitis etc. Till now more than 80 compounds, including glycosides, triterpenes, ionones, phenolics, phytosterols, organic acids, lignins, alkaloids have been isolated and identified from jute plant. The main phytochemical compounds are cardiac glycosides, corchorin, corchotoxin, helveticoside, corchoroside A and B, olitoriside, erysimoside, straphatidol, glycoside, capsularinsteroids and many other secondary metabolites. Modern studies have revealed several biological activities such as acidic polysaccharide, cardiotonic, anti-obisity, gastroprotective, antidiabetcs, antioxidant, anti-inflammatory and cytotoxic activities [**58**]. The human foreskin has potential to be used as a source of stem cells. Becer et al [**59**] investigated the apoptotic and proliferative effects of *C. olitorius* on human foreskin cells (h_nFSSCs) spheroids and found that *C. olitorius*increased proliferation of h_nFSSCs which might increase proliferation of stem cells too. It has been reported that jute halts progression of and helps manage deferent chronic diseases like tumors, obesity, diabetes, and cardiovascular diseases [**1**]. One of the major challenges facing mankind from time immemorial is disease endemic. Today, several thousands of diseases have been diagnosed, with many of them still being incurable.There are many things that can cause diseases, but microorganisms are the main cause of human, animal and plant diseases. For example, it was discovered in 19th century that microbes can cause numerous infectious diseases, like flu, measles,chickenpox, tuberculosis, whooping cough, cholera, etc. Researchers also discovered in 20thcentury that microbes contribute also to many non-infectious chronic diseases and conditions, such as some forms of cancer, coronary artery diseases, diabetes, multiple sclerosis, and chronic lung diseases. Microorganisms are of diverse kinds-pathogenic and non-pathogenic microorganisms. As non-pathogenic microbes cause no harm in their hosts, pathogenic microbes cause lots of harms (diseases and death) in their hosts. Study shows that both infectious and non-infectious human diseases are caused by pathogenic microorganisms. As the world population grossly increases, human diseases also increase.Therefore, stresses the need to research more into alternative medicine (herbal medicine), so as to augment orthodox medicine (conventional medicine). Although all parts of the plant have traditionally been used, the leaves are the most common medicinal part of the plant. Jute leaves may not be known to many, but they are a part of the jute plant that's mostly cultivated in Asia, the Middle East, and parts of Africa.

Jute leaves are used as a food source in these regions; in fact, they are said to add a distinct flavour to food and also act as thickeners in soups, stews and sauces. Jute leaves are also known as saluyot, ewedu or lalo, depending on the region they are being cultivated or cooked in. The leaves have slightly toothed edges. When harvested young, jute leaves are generally flavourful and tender; on the other hand, older leaves tend to be fibrous and woody. Scientifically known as *C. olitorius*, parts of jute plants are used in many ways. While jute stems are used to make rope, paper and a variety of other products, jute leaves are not just for culinary uses but also are known for their medicinal properties. The leaves are said to have anti-inflammatory properties, which may prevent conditions like arthritis, acne, asthma, cold, etc. Jute leaves are said to be a good source of β-carotene, which is why it is used in medicines in most parts in Africa and Middle East [**16**]. Oxidative damage contributes to the development of disease [**60, 61**]. Cooked jute appears to offer higher concentrations of these vitamins and minerals.However, that's because a cooked cup of jute leaves contains about three times the number of leaves. Gomaa et al [**62**] studied on the efficacy of *C. olitorius* leaf extract in the prevention of metabolic syndrome induced in rats by high-fat diet (HFD) and found that it is effective in preventing obesity, hyperlipidaemia, steatosis and insulin resistance. According to Gomaa et al [**62**], these actions may be mediated by inhibiting of lipase activity, TNF-α, IL-1β and leptin resistance along with increasing of adiponectin. Corchortoxin (strophanthidin) was a cardiac aglycone isolated from *C. capsularis* seeds, showed a cardio-tonic activity. These activities were similar to digitalis genus. However, jute seeds extract showed better activities than corchortoxin. Corchoroside A and B, which also isolated from other plants also showed digitalis like action [**63, 64, 65, 66, 67**]. Numerous epidemiologic literature has verified an important correlation between the consumption of phenolic compound-rich food and a decreased risk for developing cardiovascular and other diseases [**68, 69**]. Two antitumor against tumor promoter-induced Epstein-Barr virus activation were isolated from the leaves of *C. capsularis*. The antitumor-promoting activity was examined by an immuno blotting analysis. Their activecomponents were identified as phytol (3,7,11,15-tetramethyl-2-hexadecen-1-ol) and mono-galactosyldiacylglycerol (1,2-di-O-α-linolenoyl-3-O-β-D-galactopyranosyl-sn-glycerol). The content of the latter was found to vary among cultivars. The detectable amount of each active component increased by treatment of the leaves with hot water [**29**]. Brine shrimp lethality bioassay was carried out to determine the cytotoxicity of the crude methanolic extract of*C. capsularis* leaves and its fructions. Butanol extract was the most potent extract (71.14% inhibition at a concentration of 1.25mg/ml), followed by ethyl acetale (28.57% inhibition at a concentration of 1.25mg/ml) and methanol extract (14.28% inhibition at a concentration of 1.25mg/ml) [**3**]. Soykut et al [**70**] studied on the determination of in vitro anticancer and apoptotic induction effects of dichloromethane (DCM) and aqueous *C. olitorius* leaf extracts in primary (Colo-320) and metastatic (Colo-741) colon adenocarcinoma cell lines and suggested that *C. olitorius* extracts can induce apoptosis in both of these cancer cell lines while being more effective in metastatic colon adenocarcinoma cell lines.

Therefore, the extracts might have potential anticancer effects and possibility to be used as precursor to phytomedicinal colon cancer treatment as oppose to chemotherapy [**70**]. Toshio et al [**71**] isolated two antitumor promoters against tumor promoter-induced Epstein-Barr virus activation from the leaves of*C. capsularis* where they found that the detectable amount of each of these active components were increased by treatment of the leaves with hot water. Breast cancer has one of the highest mortality rates among all cancers. Soykut et al [**72**] evaluated the effectiveness of *C. olitorius* leaves extracted with ethanol plant on viability of estrogen receptor positive breast cancer cell line, MCF-7.*C. olitorius* plant contains strong phytochemicals such as quercetin and caffeoylquinic acid which exert antioxidant properties and are usually effective at reduction in cancer cell proliferation. *Corchorus* spp. leaf has long been used as a remedy in many cultures. Jute leaf products, which include the leaf juice, fried leaf, and some time whole green leaf, are used, among other reasons, as laxatives, in creams for skin care, and as atreatment for a wide range of diseases, respectively. The heterogeneous nature of jute leaf products may contribute to the diverse biological and therapeutic activities that have been observed. Variations in the composition of jute leaf can result in products with different chemical and physical properties, making the comparison of productsdifficult. The green, leafy vegetable is rich in beta-carotene for good eyesight, iron for healthy red blood cells, calcium for strong bones and teeth, and vitamin-C for smooth, clear skin, strong immune cells, and fast wound-healing. Vitamins-A, -C and -E present in jute leaf/Saluyot "spongeup" free radicals, scooping them up before they can commit cellular sabotage. Jute leaf as vegetable contains an abundance of antioxidants that have been associated with protection from chronic diseases such as heart disease, cancer, diabetes, and hypertension as well as other medical conditions. Fresh jute leaf has higher demand. Ayurvedics use the leaves for ascites, pain, piles (laxative), and tumors. Elsewhere the leaves are used for cystitis, dysuria, and fever. The cold infusion is said to restore theappetite and strength [**31**]. Leishmaniasis, the vector-borne neglected tropical disease, is the second leading infectious disease after malaria. Approximately 350 million people are affected by leishmaniasis per year in 98 countries worldwide [**73**]. Among the various clinical forms, visceral leishmaniasis is the deadliest [**74**]. Current antileishmanial drugs exhibit toxicity- and resistance-related issues. Therefore, advanced chemotherapeutic alternatives are in demand, and currently, plant sources are considered preferable choices. Pramanik et al [**74**] reported that the chloroform extract of *C. capsularis*leaves exhibits a significant effect where leaf-derived β-sitosterol competitively inhibited Leishmaniadonov anitrypanothione reductase. *C. capsularis* leaf-derived β-sitosterol (β-sitosterol CCL) as a novel antileishmanial agent that competitively inhibits Leishmaniadonov anitrypanothione reductase.Overall, the antiparasitic efficiency and the ability of this phytosterol to block one of the most crucial parasitic enzymes emphasize its proficiency as a strong candidate for treatment of visceral leishmaniasis [**74**]. İşeri et al [**75]** investigated cytotoxic and genotoxic effects of leaf extracts and seed extracts of the *C. olitorius* on the multiple myeloma-derived ARH-77 cells where the extracts were also evaluated for their total phenol content and free radical scavenging activity and reported the high cytotoxic potential of *C. olitorius* seed extracts and the genotoxic potential of leaf extracts and seed extracts. Dewanjee et al [**76**] evaluated the protective effect of aqueous extract of *C. olitorius* leaves against $CdCl_2$ intoxication and their results exhibited a significantly increased intracellular Cd accumulation, oxidative stress and DNA fragmentation in the organs. Moreover, the haematological parameters were significantly altered in the $CdCl_2$ treatedand could significantly restore the biochemical, antioxidant and haematological parameters near to the normal status. Their histological studies of the organs supported the protective role of jute leaves, in addition the presence of substantial quantity of phenolic compounds and flavonoids in extract may be responsible for overall protective effect. Jute leaves also contain lycopene, an antioxidant that protects our cells against oxidative damage, which increases your disease risk. Their lycopene content can vary by the preparation method. For example, cooked jute leaves and older jute leaves have higher levels [**77**]. Jute leaves may have an anti-inflammatory effect on the liver. A study examined rats who ate jute leaves daily for 30 days. At the end of the study, the rats had improved liver antioxidant statuses [**78**]. Groundwater arsenic contamination in Bangladesh and its adjoining part of West Bengal (India) is reported to be the biggest arsenic calamity in the world in terms of the affected population. Tossa jute (*C. olitorius*) is a popular crop of this arsenic prone population. Das et al [**79**] evaluated the protective effect of aqueous extract of *C. olitorius* leaves against sodium arsenite (NaAsO(2)) induced cardiotoxicity in experimental rats and concluded that the treatment with aqueous extract of *C. olitorius* prior to arsenic intoxication has significant protecting effect against arsenic-induced myocardial injury. On the other hand, Das et al [**80**] studied to evaluate the protective effect of aqueous extract of *C. olitorius* leaves against sodium arsenite-induced toxicity in experimental rats and concluded that the treatment with aqueous extract of *C. olitorius* prior to arsenic intoxication has significant role in protecting animals from arsenic-induced hepatic and renal toxicity. Again, Das et al [**81**] suggested that treatment with aqueous extract of*C. olitorius* prior to arsenic intoxication has a significant role in protecting animals from arsenic-induced toxicity too. Rice accumulatesthe arsenic in different parts during irrigation with arsenic contaminated water and this arsenic are accumulated in different parts of rice which increased dose-dependently. Hence, for irrigation purpose arsenic contaminated water cannot be used. Furthermore, arsenic contaminated rice induced several toxicities in animal model, most of which could be minimized with the food supplementation of*Corchorus olitorius* leaves. Therefore,*C. olitorius*can be used as a potential food supplement to the affected people of arsenic prone zone to ensure the food security [**82**]. Yan et al [**83**] investigated the effect of five fractions (crude phenolic extracts using 80 per cent aqueous acetone of whole plant, leaf, stem, washed leaf (WL) and dried water washing material (WW) on the regulation of inflammatory responses in lipopolysaccharide (LPS)-stimulated J774A.1 macrophage cells and found that Linoleic acid autoxidation inhibitions on all fractions were higher than that on α-tocopherol. Moreover, all fractions diminished LPS-induced protein expression of inducible nitric oxide synthase (iNOS) and cyclooxygenase 2 (COX-2). Nitric oxide (NO) and prostaglandin E2 (PGE(2)), downstream products, were also suppressed in dose-dependent manners, except for WL and WW. It was also found that oxidative modification and loss of leaf phenolics after kneading and washing greatly affected DPPH radical scavenging and inflammatory responses [**83**]. The mosquitocidal activities of *C. capsularis* against a common malarial vector, Anopheles stephensi and a dengue vector Aedes aegypti was studied where Elangovan et al [**84**] referred to the possible utilization of *C. capsularis*to control mosquito menace to a greater extent. However, On the other hand, in studying of the role of jute leaf (*C. capsularis*) phytochemicals on feeding, growth and reproduction of *Diacrisiacasignetum*Kollar (Lepidoptera: Arctiidae), it appeared that the larval and post larval developmental duration was shorter on mature jute leaf fed insects whereas adult longevity was higher in it ($P< 0.05$) relative to young and senescent leaf fed insects. Fecundity of *D. casignetum*was also highest on mature leaves followed by young and senescent leaves. The growth and development of*D. casignetum* were related to the nutrient content relative to the secondary metabolites of these three types of jute leaves. Higher levels of nutritional factors (total carbohydrates, proteins, lipids, nitrogen and amino acids including water content) and lower levels of anti-nutritional factors (secondary metabolites) in mature jute leaves have influenced lower developmental time along with higher growth rate, fecundity and accumulated survivability of *D. casignetum* than the young and senescent leaves [**84, 85**].

The use of functional foods and their bioactive constituents have been considered as a new approach in the prevention and management of type-2 diabetes and its complications. Saliu et al [**86**] evaluated the antidiabetic potentials of jute leaf (*C. olitorius*) on low dose streptozotocin/high fat diet induced diabetic rats and found that jute leaf could be a good source of functional food for dietary intervention in the management of type-2diabetes and its associated complications. Oboh et al [**87**] reported the inhibitory action of polyphenol-rich extracts (free and bound) of *C. olitorius*on α-amylase, α-glucosidase and angiotensin I converting enzyme (ACE), as well as to identify the phenolic compound responsible for these activities. From their study it was found that the enzyme inhibitory activity of *C. olitorius* extracts may be attributed to the presence of caffeic acid, chlorogenic acid and isorhamnetin, which justify the use of jute leaf in folklore for the management of diabetes and hypertension. Saliu et al [**78**] studied on impaired liver function, which is associated with decreased hepatic δ-aminolevulinic acid dehydratase (δ-ALAD) activity, in diabetes mellitus and described the effect of dietary jute leaf (*C. olitorius*) on hepatic δ-ALAD activity in high-fat fed combined with low-dose streptozotocin administered diabetic rats. Saliu et al [**78**] suggested that restoration of hepatic δ-ALAD activity, modulation of hepatic function biomarkers, and increase in antioxidant status could be possible underlying events mediating the hepatoprotective effect of jute leaf in diabetic conditions. Shariare et al [**88**] reported that Phytol, a pharmacologically active compound present in *C. olitorius* leaf exhibit a range of activity including anti-inflammatory, antioxidant, anticancer, hepatoprotective etc. In-vivo anti-inflammatory study showed significant activity for *C. olitorius*leaf extract against carrageenan induced paw edema, which is significantly increased while delivered throughliposomes. In-vitro cancer cell line study data suggests that liposomal delivery of phytol was more active at lower concentration compared to pure phytol, for specific cell lines [**88**]. *C. olitorius* is reportedly used in ethno-medicine to arrest threatened miscarriage and other conditions associated with excessive uterine contractions. The plant is also used as a purgative, demulscent and an anti-inflammatory agent [**89**]. According to Orieke et al [**89**] *C. olitorius* leaf extract (COLE) may be a good tocolytic, anti-diarrheal and anti-inflammatory agent and offers hope of new drug discovery for such uses. Owoyele et al [**90**] provided information about the antipyretic and anti-inflammatory effects of *C. olitorius* root and showed that *C. olitorius* root is another good source of phytomedicine that can be used effectively to treat inflammation and pyrexia that accompany some diseases.Several studies have reported on *C. olitorius* regarding various pharmacological properties, such as antioxidant, anti-inflammatory, analgesic, and anticancer properties. Tosoc et al [**91**] studied to evaluate the anticancer properties, such as the antiproliferative, anti-angiogenic, and antitumor activities, of the *C. olitorius* aqueous extract and its bioactive compounds, chlorogenic acid (CGA) and isoquercetin (IQ), against human melanoma (A-375), gastric cancer (AGS), and pancreatic cancer (SUIT-2), using in vitro and in vivo biological assays where they concluded that *C. olitorius* aqueous extract has promising anticancer effects on various types of human cancer cells. Moreover, they inferred that *C. olitorius* aqueous extract and its compounds are potential nutraceutical products that could be used for cancer treatment. Recently, *C. olitorius* has been approved for its antitumor activity (Li et al 2012). According to Li et al [**92**], ethanol extract of *C. olitorius* might be effective against hepatocellular carcinoma through induction of apoptosis via mitochondria-dependent pathway. *C. olitorius* is an African leafy vegetable of high nutritional interest. Guzzetti et al [**93**] assessed its agricultural suitability to sustainable cultivation conditions and its potential benefits for human nutrition, its phytochemical content in response to conservation agriculture practices [i.e., no-tillage (NT) and cover crop maintenance] and low water regime were evaluated and compared with response under conventional agriculture management. Their research showed that *C. olitorius* can be a promising crop for improving both agricultural sustainability and health benefits due to the great amount of antioxidant compounds in leaves, whose occurrence is not altered by stressful farming conditions. Azuma et al [**14**] identified six phenolic antioxidative compounds [5-caffeoylquinic acid (chlorogenic acid), 3,5-dicaffeoylquinic acid, quercetin 3-galactoside, quercetin 3-glucoside, quercetin 3-(6-malonylglucoside), and quercetin 3-(6-malonylgalactoside) (tentative)], ascorbic acid and alpha-tocopherol from the leaves of *C. olitorius* by NMR and FAB-MS. Their results showed that 5-caffeoylquinic acid was a predominant phenolic antioxidant in *C. olitorius* leaves [**14**].

2.2.4.2.6. Antioxidants activity Antioxidants also turn free radicals into waste by-products, which are eliminated from the body. Consumption of antioxidant-enriched fruits and vegetables is known to lower the risk of several diseases caused by free radicals [**94**]. Such health benefits are mainly due to the presence of phytochemicals such as polyphenols, carotenoids, and vitamin-E and -C [**95**]. Oboh et al [**96**] studied a comparative study of the antioxidant properties of hydrophilic extract and lipophilic extract constituents of the leafy vegetable and concluded that the additive/synergistic antioxidant activities of the hydrophilic and lipophilic constituents may contribute to the medicinal properties of *C. olitorius* leaf. The protection fruits and vegetables provide against diseases are attributed to the various antioxidants they contain **[97]**. Antioxidantsare compounds that inhibit or delay the oxidation of other molecules by inhibiting the initiation or propagationof oxidizing chain reactions [**98**]. They neutralize free radical reactive species that are generated endogenously through aerobic metabolism [**99**]. Antioxidants are substances that prevent and stabilize the damage caused by free radicals by supplying electrons from antioxidants to these damage cells. Antioxidantsmay be defined as complexdetermined compounds that function asdefensive shields against several diseases [**100**]. Phenolic compounds are an important group of plant-based biologically active compounds that strengthen the organism and prevent disease [**101,102**]. Plant polyphenols are secondary metabolites characterized by one or more hydroxyl groups binding to one or more aromatic rings [**103**]. Phenolic compounds have a particularly strong antioxidant effect [**104, 105, 106**]. Omenna and Ojo [**107**] evaluated the comparative phytochemical composition of kenaf and jute leaves where they infer that kenaf and jute leaves are rich in phytochemicals and antioxidants. Alexopoulou et al [**108**] reported that kenaf leaf was applied to Guinea worms and the stem bark has been used for anaemia in Africa. In addition, Jaihyunk et al [**109**] reported the presence of phyto compounds in the hexane extracts of the different parts of the kenaf plant by GC-MS analysis. Copper cause oxidative damage in plant cells and plant extracts are the sources of free radical scavengers. İşeri et al [**110**] tested the hypothesis that whether *C. olitorius*and*Urtica pilulifera*(Roman nettle) seed extract treatments of germinated seeds affect copper induced oxidative and genotoxic damage or antioxidant response in tomato where they found that jute seed extract contained salicylic acid and quercetin which can be correlated with the evoked effects. In wound healing, antioxidants counter the excess proteases and reactive oxygen species (ROS) often formed by neutrophil accumulation in the wounded area and protect protease inhibitors from oxidative damage. Fibroblasts and other cells may be killed by excess ROS and skin lipids will be made less flexible.Antioxidants substances reduce the possibility of these adverse events occurring hence appear tobe important in the successful treatment of wounds [**111, 112**]. Compounds with high radical-scavenging capacity have been shown to facilitate wound-healing. Vya et al [**113**] reported that the methanol extract of *C. olitorius* leaves exhibited a high degree of woundhealing activity and the antioxidant activity. Moreover, according to Vya et al [**113**], *C. olitorius* displayed significant free-radical-scavenging activity in a concentration dependant manner in in-vitro assays. Also, the powder ointment and the water extract of *C. olitorius* have been shown to have a wound-healing property. The traditional uses of this plant to treat wound has therefore been confirmed. The free radical scavenging properties of *C. capsularis* found in Malaysia was studied [**40**]. The air-dried leaves of the plant (20g) were soaked in distilled water (1:20;$^{w}/_{v}$) for 72h at room temperature. The collected supernatants were tested for the free radical scavenging activity against the DPPH and superoxide anion radical scavenging assays. The extract showed remarkable antioxidant activity in both assays with the percentage of inhibition nearly 90% [**40**]. The crude methanolic extract of *C. capsularis* (leaves) and its fructions (5-25μg/μl), were tested for the free radical scavenging activity against the DPPH and superoxide anion radical scavenging assays. Extracts were found to show remarkable antioxidant activity in both assays with the percentage of inhibition. Hexan extract caused 65.44-97.43 % inhibition and appeared the most potent antioxidant extract, followed by butanol, methanol and ethyl acetate extracts [**3**]. Sadat et al [**114**] carried out a study to evaluate the in vitro preliminary phytochemical analysis and antioxidant activity of methanolic leaf extracts of *C. olitorius* and revealed that the jute leaf possesses different phytonutrients and exhibited 1,1-diphenyl 2-picrylhydrazyl radical scavenging activity and therefore, may be used for therapeutic purposes. Various phytochemical, mineral, and antioxidant potency properties of 30 genotypes belonging to *C. capsularis* and *C. olitorius* were evaluated by Biswas et al [**1**]. Ali et al [**32**] investigated on the antioxidant activities and HPLC profiling of polyphenolic compounds in ethanol extract of *C. olitorius* and *C. capsularis* leaves where they evaluated the total antioxidant capacity by phosphomolybdenum method and identification and quantification of polyphenolic compounds by HPLC-DAD system. According to their results, eight polyphenolic compounds were found in the*C. capsularis* leaves but *C. olitorius* leaves contain six polyphenolic compounds. Therefore it could be suggested that *C. olitorius* and *C. capsularis* leaves contain a significant amount of several polyphenolic compounds that could be used as a natural antioxidant for functional foods [40]. Leaves of *C. olitorius* have a large quantity of antioxidants compounds connected with various biological properties, which include diuretic, analgesic, antipyretic, antimicrobial activities, antitumor [**53**], phenolic antioxidative compounds [**14**], hypoglycemic and gastroprotective [**115**]. Jute leaves are rich in Vitamins-E, -A and -C which are potent antioxidants that protect the cells from free radicals and can help to improve vision and fertility. It is also found to be rich in antioxidant compounds such as caffeoylquinic acid, quercetin glycosides, vitamin-E and omega-3 fatty acids [**14, 13, 16, 116**].

2.2.4.2.7. Anti-inflammatory, analgesic, cytotoxic and antipyretic effects The antinociceptive and anti-inflammatory properties of *C. capsularis* leaves chloroform extract were investigated in experimental animal models. The antinociceptive activity was measured using the writhing, hot plate and formalin tests, while the anti-inflammatory activity was measured using the carrageenan-induced paw edema test. The extract was used in the doses of 20, 100 and 200mg/kg. It was administered subcutaneously, 30min prior to subjection to the respective assays. The extract was found to exhibit significant ($p<0.05$) antinociceptive and anti-inflammatory activities Zakaria et al [**40**]. Dietary phenolics are known for their potent antioxidant and anti-inflammatory activities, making them promising candidates for protection against neuroinflammation and neurodegeneration. Wagdy et al [**117**] investigated the hydroalcohol extract of Egyptian species of *C. olitorius* leaves for its neuroprotective effects in a lipopolysaccharide-induced neuroinflammatory mouse model where they showed the promising effects of *C. olitorius*leaves in limiting neurodegeneration and cognitive impairment caused by neuroinflammation and glial cell activation. Their findings can be use in the modulation of neuroinflammatory pathways improving the neuroinflammation-related neurodegeneration and cognitive decline. This makes *C. olitorius* leaves a promising candidate as a nutraceutical supplement to be used against neuroinflammation-related disorders [**117**]. The antinociceptive, anti-inflammatory and antipyretic properties of an aqueous extract of *C. capsularis* leaves were studied in experimental animals. The antinociceptive activity was measured using the abdominal constriction, hot plate and formalin tests, while, the anti-inflammatory and antipyretic activities were measured using the carrageenan-induced paw edema and brewer's yeast-induced pyrexia tests, respectively. The extract was used as 11.57, 57.85, and 115.7mg/kg, it was administered subcutaneously, 30min prior to subjection to the mentioned assays. The extract was found to exhibit significant antinociceptive, anti-inflammatory and anti-pyretic activities in a dosage-independent manner [**53**]. Özlem et al [**118**] investigated and evaluated the cytotoxic and genotoxic effects, total phenol content and free radical scavenging activity of leaf extracts and seed extracts of the *C. olitorius* on the multiple myeloma-derived ARH-77 cells and it was revealed that the extracts induced genotoxic damage on ARH-77 cells. They demonstrated the high cytotoxic potential of *C. olitorius* seed extracts and the genotoxic potential of leaf extracts and seed extracts. It can be concluded from their study that In vitro cytotoxic effect and genotoxicty of leaf extracts may be attributed to pro-oxidant effect of phenol compounds, whereas cytotoxic potential of seed extracts may be related to direct effect of bioactive constituents [**118**].

2.2.4.2.8. Antimicrobial effect Senu et al [**119**] evaluated the antimicrobial effect of the ethanol extracts of *C. olitorius* leaf at 150, 100, 50, 25mg/mL on eight organisms and found that at 100mg/mL and below, all the bacteria were resistant to the extract except Shigelladysenteria which was susceptible at the concentration of 100mg/mL, but at the concentration of 150mg/ml all the bacteria were susceptible except *Bacillus subtilis*. Rume [**3**] used disc diffusion method to determine the antibacterial and antifungal activity of the crude methanolic extract of *C. capsularis* leaves and its functions against Gram positive bacteria (*Bacillus subtilis, Staphylococcus aureus*, β-hemolytic *Streptococcus, Bacillus cereus* and *Streptococcus pyrpgen*), Gram negative bacteria (*Shigella boydii, Salmonella typhi, E. coli, Klebsiella* sp. and*Vibrio mimicus*), yeast and fungi (*Candida albicans, Saccharomyces cerevisiae* and *Bacillus megaterium*). *C. capsularis* extracts possessed antifungal and anti-yeast activity, and N-hexane fraction of methanolic extract of leaves of *C. capsularis* showed the highest activities against Gram positive, Gram negative bacteria and fungi with a zone of inhibition 0.9-1.5mm, followed by hexane extract [**3, 120**]. Yakoub et al [**121**] analyzed the antioxidant and antibacterial activities of *C. olitorius* extracts. Antimicrobial test showed also that increasing concentrations of phenolic extracts increased the clear diameter zone around the tested bacteria and that ethanolic extract exhibited the best antibacterial activity against *Staphylocaccusaureus, Micrococcus luteus, Bacillus cereus, Escherichia coli, Klebsiella pneumoniae, Salmonella enterica, Salmonella typhi, and Enterobacter* sp. [**121**] which suggest that ethanolic extract might be used as a natural antioxidant and antimicrobial for functional foods. Hayder and Enas [**122**] investigated the chemical constituents, pharmacological activity and antiviral activity of the *C. olitorius* against measles virus, where they stated the significant antiviral activityof the aqueous layer of *C. olitorius* leaves against measles virus which was due to the presence of phenols and flavonoids in the aqueous layer of *C. olitorius* leaves.

2.2.4.2.9. Insecticidal effect and pesticidal use The mosquitocidal activities of *C. capsularis* against a common malarial vector, Anopheles stephensi and a dengue vector *Aedes aegypti* were studied. The larvicidal activity exerted by ethyl acetate was more prominent than acetone and methanol extracts in all concentrations tested against Ae. aegypti larvae. Evaluation of the lethal concentration values (LC_{50} and LC_{90}) of acetone, ethyl acetate and methanol extract of the plant against *An. stephensi* and *Ae. aegypti* revealed that LC_{50} of 197.34ppm and LC_{90} of 358.59ppm was recorded for acetone extract against the *An. stephensi*; furthermore, the larvae of *Ae.*

*aegypti*showed the LC_{50} and LC_{90} values of 222.45 and 383.06ppm respectively, with the treatment with the acetone extract of *C. capsularis*. Minimum LC_{50} values were observed among the experimental larval groups treated with methanol extract of *C. capsularis* were 176.19ppm and 182.06ppm against *An. stephensi* and *Ae. Aegypti*, respectively. With regard to the ovicidal activity of acetone, ethyl acetate and methanol extract, it was apparent that 300-450ppm concentrations resulted with no hatchability on *An. stephensi* and 375-450ppm concentrations in*Ae. aegypti*. The authors referred to the possible utilization of*C. capsularis* to control mosquito menace to a greater extent [**84**]. The efficacy of emulsified petroleum ether extract of Corchorus capsularis seed was studied against three stored product pests (*Callosobruchus chinensis*, *Sitophilusoryzae* and *Tribolium castaneum* Herbst) in adult phase. The residual film technique method was conducted to determine the LC_{50} value of the mentioned plant extract against three stored product pests. LD_{50}(µg /cm) of *C. capsularis* against *C. chinensis* was 74.26 (50.26-109.74) after 24h and 6.67 (0.49-90.07) after 48h. LD_{50} against *S. oryzae* was 84.61 (61.98-115.50) after 24 h and 32.87 (16.03-67.39) after 48h. While, LD50 against *T. castaneum* was 547.08 (477.38-626.97) after 24h and 452.51 (380.30-538.42) after 48h [**123**]. However, On the other hand, in studying of the role of *C. capsularis* leaf phytochemicals on feeding, growth and reproduction of *Diacrisiacasignetum*Kollar (Lepidoptera:Arctiidae), it appeared that the larval and post larval developmental duration was shorter on mature jute leaf fed insects whereas adult longevity was higher in it ($P< 0.05$) relative to young and senescent leaf fed insects. Fecundity of *D. casignetum*was also highest on mature leaves followed by young and senescent leaves. The growth and development of*D. casignetum* were related to the nutrient content relative to the secondary metabolites of these three types of jute leaves. Higher levels of nutritional factors (total carbohydrates, proteins, lipids, nitrogen and amino acids including water content) and lower levels of anti-nutritional factors (secondary metabolites) in mature jute leaves have influenced lower developmental time along with higher growth rate, fecundity and accumulated survivability of *D. casignetum* than the young and senescent leaves [**84**]. Jute leaves contain Hydrogen cyanide (HCN) and several cardiac glycosides. Negm et al [**65**] report the LD_{50} of issue extracts to mice. The lethal dose of Corchoroside A to cats is 0.053-0.0768mg/kg and Corchoroside B 0.059-0.1413, but some authors say that Corchoroside A is twice as active as Corchoroside B [**124**].Anthracnose spots caused by *Colletotrichumgloeosporioides* may infect 50-90% of a jute population, but spraying with copper oxychloride at 0.5% strength checked the spread, holding it to 5-10%. This species was badly infested by 3 species of weevils (*Myllocerus*spp.) while *C. capsularis* was unaffected. The semilooper (*Anomissabulifera*) may stunt the growth, reducing fiber yields by ca 13-32 %. The yellow mite (*Polyphagotarsonemuslatus*) may also reduce yields [**124**].

2.2.4.2.10. Source of nanoparticles The green synthesis of nanoparticles has attracted many researchers because of the high demand to produce clean, nontoxic chemicals, environmentally benign solvents, and renewable materials. Moreover, this methodology improves the biocompatibility of the obtained NPs [**79, 125, 126, 127**]. For instance, many successful attempts have been reported in the literature, which achieve the synthesis of nanoparticles using biological systems, such as yeast, fungi, and bacteria [**79, 125, 126, 127, 128**]. Recently, the use of plant extracts has proven to be one of the most successful techniques for growing nanoparticles. Most of these studies employed broths, obtained from boiling fresh plant leaves [**66,129, 130, 131, 132, 133, 134**]. So far, it has been established that the antioxidant components of the studied plant extracts are responsible for the reduction of metal salts, leading to the growth and stabilization of the NPs [**129, 130, 66, 127,131, 132, 133, 134**]. The plant extracts are generally mixtures containing a variety of compounds, including active component(s) [**126, 127,135**]. In spite of the fact that each active constituent in herbal extracts claims efficacy, the combined components may reciprocally affect their respective pharmacokinetic behavior, thus hampering the extract's safety and efficiency [**135, 136, 137, 138**]. The eco-friendly process for the synthesis of nanoparticles like-silver nanoparticles (AgNP) and gold nanoparticles (AuNP) are now focuses on mechanism of the antibacterial activity of AgNPs and the anticancer activity of AuNPs. Biomolecules in the plant extract are involved in reduction of metal ions to nanoparticle in a one-step and eco-friendly synthesis process. Natural plant extracts contain wide range of metabolites including carbohydrates, alkaloids, terpenoids, phenolic compounds, and enzymes. A variety of plant species and plant parts have been successfully extracted and utilized for AgNP and AuNP syntheses. Green-synthesized nanoparticles eliminate the need for a stabilizing and capping agent and show shape and size-dependent biological activities. Nanoparticles are important in the field of pharmaceuticals for their strong antibacterial and anticancer activity [**27, 139,140,141**]. Gold nanoparticles (AuNPs) are extremely promising objects for solving a wide range of biomedical problems. The gold nanoparticles production by biological method ("green synthesis") is eco-friendly and allows minimization of the amount of harmful chemical and toxic byproducts [**142**]. A facile bottom-up "green" synthetic route of gold nanoparticles (Au NPs) is described, using a leaf extract of *C. olitorius* as a reducing and stabilizing agent [**127**]. Ismail et al [**127**] reported a one-step green synthesis of gold nanostructures using naturally occurring biodegradable plant-based surfactants from *C. olitorius* leaves, without any special reducing or capping agents.

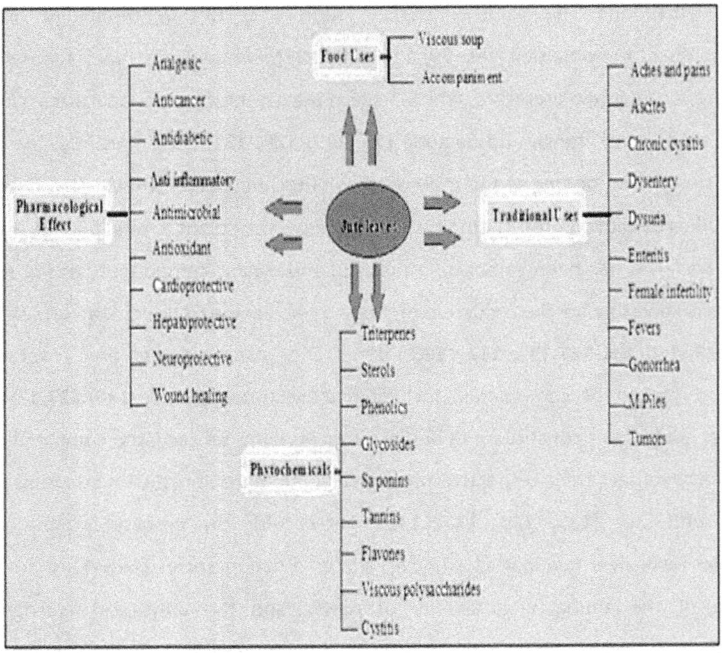

Figure 2.2.4.2.1 Different uses of jute leaf. [Zakaria and Shuranjan 2022]

2.2.4.2.11. Conclusion

Demand for medicinal plants is increasing in both developed and developing countries due to the growing recognition of natural products being equally effective, safe, non-narcotic, and affordable and has no side effects. Biologically active pure compound is better than crude extract. However, to get the overall view regarding the phytochemical composition and biological activities of any plant, it is important to pick the most potential part by investigating several parts (leaf, stem bark, root bark, fruits, flowers etc.) of that plant. Secondary metabolite molecules in plants food are attributed withvarious biological activities. Various findings demonstrate that leaves of *C. capsularis* and *C. olitorius* with the high level of phytochemical constituents include flavonoids, polyphenols, tannins, and saponins that possess strong radical scavenging activity and antioxidant power. Therefore, this jute leaf could be a promising potential as an outstanding source of natural antioxidants in food, pharmaceutical, and cosmetics industries. Researchers studied the phytochemical and antioxidant activity of methanolic extract of leaves of C. olitorius and find out the presence of different secondary metabolites and antioxidants. The presence of these bioactive compounds in jute leaf establishes it as a potential source of a natural therapeutic agent that can be used against different infectious and other diseases. These extracts might have potential anticancer effects and possibility to be used as precursor to phytomedicinal colon cancer treatment. *C. olitorious* is easily available and economically cheap and therefore can be used for the medicinal purposes. Finally, investigations on the isolation and identification of antioxidant components in this particular plant will lead us to know the chemical entities with potential for clinical use. The *C. olitorius* extracts may serve as a substitute for chemical antioxidant substances and hence might have the potential to serve as bio-preservatives too. *C. capsularis* can be taken as alternative sources for other constituents applicable tothe nutrition, medicine and cosmetics. The phytochemical and antioxidant properties of methanolic extract of leaves of *C. olitorious* indicated the presence of different secondary metabolites and antioxidants. According to Ahmed [**6**], jute leaf has diverse pharmacological activities due to presence of polysaccharides and phenolic compounds which would be beneficial to popularize the utilization of jute leaf as functional food ingredient in food formulations. The presence of these bioactive compounds in jute leaf establishes it as a potential source of a natural therapeutic agent that can be used against different infectious and other diseases. *C. olitorious* is easily available and economically cheap and therefore can be used for the medicinal purposes.

2.2.4.2.12 References

[1] Biswas A, Dey S, Li D, Liu Y, Zhang J, Huang S, Pan G and Deng Y. Comparison of phytochemical profile, mineral content, and In vitro antioxidant activities of Corchorus capsularis and Corchorus olitorius leaf extracts from different populations. J Food Quality. Article ID 2931097, 2020; 14 pages.

[2] Islam MM. Rahman M. In: Hand book on agricultural technologies of jute, kenaf and mesta crops. Bangladesh Jute Research Institute, Manik Mia Avenue, Dhaka-1207, Bangladesh, 2008; 92.

[3] Rume JM. Phytochemical, antimicrobial and biological investigations of methanolic extract of leaves of Corchorus capsularis. Thesis for bachelor degree of pharmacy, 2010; East West University.

[4] Kirtikar KR. Basu BD. Medicinal Plants of India. New Delhi: Indian Council of Medical Research. 1975; Vol. 4.

[5] Vorpur.com, Bangladesh's Leading Online Health Food Store Based On Dhaka, http://www.vorpur.com/vegetable-organic-jute-vegetables-spinach-leaves-argaanik-pat-shak/.

[6] Ahmed Z, Banu H, Akhter F. Izumori K. A simple method for D-xylose extraction from jute stick and rice husk. Online J Biol Sci. 2001; 1: 001-1004.

[7] Holm LG. A Geographical Atlas of World Weeds. 1979; New York: Wiley.

[8] Mahato SB. Subodh K. Cardiac Glycosides from Corchorus olitorius. 1989; 2065-8.

[9] Goda Y. Sakai S. Nakamura T. Akiyama HTM. Identification and Analyses of Main Cardiac Glycosides Seeds and Their Acute Oral Toxicity in Corchorus to Mice olitor. 1998; J Food Hyg Soc Japan.

[10] Yoshikawa M. Murakami T. Shimada H. Fukada N. Matsuda H. Sashida Y. et al. Corchorusosides A, B, C, D, and E, new cardiotonic oligoglycosides from the seeds of Corchorus olitorius L. (Moroheiya). Heterocycles. 1998; 5(48): 869-73.

[11] Nakamura T. Goda Y. Sakai S. Kondo K. Akiyama H. Toyoda M. Cardenolide glycosides from seeds of Corchorus olitorius. Phytochemistry. 1998; 49(7): 2097-101.

[12] Masayuki V. Shimada H. Saka M. Yoshizumi S. Yamaharaa J. Matsuda H. Absolute stereostructures of corchoionosides A,B and C, histamine release inhibitors from the leaves of vietnamese Corchorus olitorius l. (tiliaceae). Chem Pharm Bull. 1997; 45(3): 464-9.

[13] Handoussa H. Hanafi R. Eddiasty I. El-Gendy M. El Khatib A. Linscheid M. et al. Anti-inflammatory and cytotoxic activities of dietary phenolics isolated from Corchorus olitorius and Vitis vinifera. J Funct Foods. 2013; 5(3): 1204-16.

[14] Azuma K. Nakayama M. Koshioka M. Ippoushi K. Yama-guchi Y. Kohata K, et al. Phenolic antioxidants from the leaves of Corchorus olitorius L. J. Agri. Food Chem. 1999; 47(10): 3963–6.

[15] Hiroshi K. Seiji T. Yasutoshi YMS. Cconstituents of Corchorus olitorius. Natural Medicine. 1994; 48(3): 213-214.

[16] Zeghichi S. Kallithraka S. Simopoulos AP. Nutritional composition of molokhia (Corchorus olitorius) and stamnagathi (Cichorium spinosum). World Rev Nutr Diet. 2003; 91:1-21.

[17] Mibei EK. Ojijo NKO. Karanja SM. Kinyua JK. Phytochemical and antioxidant analysis of methanolic extracts of four african indegenous leafy vegetables. Ann Food Sci Technol. 2012; 13(1): 37-42.

[18] Qureshi AK. Chemical constituents of Corchorus capsularis and C.olitorius (Jute Plant), Part IV: Isolation of Corosolic Acid, Ursolic Acid and corosin and Correlation of Corosin with Tormentic Acid. 2013; 65(4): 1473-8.

[19] Manzoor-I-Khuda M. Habermehl G. Chemical constituents of Corchorus capsularis and C. olitorius (jute plant), Part IV isolation of corosolic acid, ursolic acid and corosin and correlation of corosin with tormentic acid. Z Naturforsch B. 1979; 34: 1320-5.

[20] Hasan HT. Kadhim EJ. Phytochemical investigation of Corchorus olitorius L. leaves cultivated in Iraq and it's in vitro antiviral activity. Iraqi J Pharm Sci. 2018; 27: 115-22.

[21] Ragasa CY. Vivar JLA. Tan MCS. Shen CC. Chemical constituents of Corchorus olitorius L. Int J Pharmacogn Phytochem Res. 2016; 8(12): 2085-9.

[22] Ademiluyi AO. Oboh G. Aragbaiye FP. Oyeleye SI. Ogunsuyi OB. Antioxidant properties and in vitro α-amylase and α-glucosidase inhibitory properties of phenolics constituents from different varieties of Corchorus spp. J Taibah Univ Med Sci 2015; 10: 278-87.

[23] Ben YAR. Abdehedi O. Jridi M. Elfalleh W. Nasri M. Ferchichi A. Flavonoids, phenols, antioxidant, and antimicrobial activities in various extracts from Tossa jute leave (Corchorus olitorus). Ind Crops Prod. 2018; 118: 206-13.

[24] Mekhael MKG. Hassan A. Hanna A. Simon A. Tóth G. Duddeck H. Phytochemical investigation of Corchorus olitorius and Corchorus capsularis (family Tiliaceae) that grow in Egypt. Egypt Pharm J 2019; 18: 123-34.

[25] Kohda H. Tanaka S. Yamaoka Y. Moringa S. Ohara Y. Constituents of Corchorus olitorius L. Natu Med. 1994; 48: 213-4.

[26] Nduka AG. Oluwatoyin AO. Temitope AO. Odutola O. Hypoglycemic and hypolipidemic effects of Corchorus olitorius leaves as a food supplement on rats with alloxan-induced diabet. Ukrain Biochem J. 2020; 92: 63-9.

[27] Patil DK. Jain AP. Extraction, qualitative and quantitative determination of secondary metabolites of Corchorus olitorius. J Drug Deliv Ther. 2019; 9(1-s): 252-255.

[28] Driss D. Kaoubaa M. Ben MR. Kallel F. Eddine AB. Ellouz CS. Antioxidant, antimutagenic and cytotoxic properties of essential oil from Corchorus olitorius L. flowers and Leaf. Free Radic Antioxid. 2016; 6: 34-43.

[29] Furumoto T. Wang R. Okazaki K. Hasan AF. Ali MI. Kondo A. et al. Antitumor promoters in leaves of jute (Corchorus capsularis and Corchorus olitorius). Food Sci Technol Res. 2002; 8: 239-43.

[30] Khan MSY. Bano S. Javed K. Mueed MA. A comprehensive review on the chemistry and pharmacology of Corchorus species- a source of cardiac glycosides, triterpenoids, ionones, flavonoids, coumarins, steroids and some other compounds. J. Sci. Indus. Res. 2006; 65: 283–298.

[31] Islam MT. Freitas RM. Sultana I. Mahmood A. Hossain JA. Homa Z. Chowdhury MMU. A Comprehensive Review of Corchorus Capsularis: A Source of Nutrition, Essential Phytoconstituents and Biological Activities. J Biomed Pharma Res. 2013; 2 (1): 1-8.

[32] Ali MM. Ahmed KS. Hossain H. Roy B. Rokeya B. Rahman MT. Jahan IA. Rahman MM. Total antioxidant capacity and profiling of polyphenolic compounds in jute leaves by HPLC-DAD. Food Res. 2021; 5 (1): 343 –348.

[33] Adeniyi SA. Ehiagbonare JE. Nwangwu SCO. Nutritional evaluation of some staple leafy vegetables in Southern Nigeria. Int J Agri Food Sci. 2012; 2(2): 37-43.

[34] Barku VYA. Boye A. Quansah N. Antioxidant and wound healing on the extrats of Corchorus olitorius leaf. World Essays J. 2013; 1(3): 67-73.

[35] Elangovan A. Dhanasekaran S. Anandan A. Krishnappa K. Gokulakrishnan J. Elumalai K. Mosquitocidal activities of Corchorus capsularis L (Malvaceae) against a common malaria vector, Anopheles stephensi (Liston) and a dengue vector Aedes aegypti (L) (Diptera: Culicidae). Int J Recent Sci Res. 2012; 3(6): 564-568.

[36] Hasan CM. Islam A. Ahmed M. Ahmed M. Waterman PG. Capsugenin, a dammarane triterpene from Corchorus capsularis. Phytochemistry. 1984; 23(11): 2583- 87.

[37] Ramadevi D. Ganapaty S. Phytochemical examination of Corchorus capsularis roots. Int J Pharmacognosy Phytochem Res. 2013; 5 (3): 173-76.

[38] Tabassum T. Extraction, identification and estimation of caffeine and catechin from Corchorus capsularis leaves extract. Thesis of a bachelor degree in pharmacy, 2009; East West University.

[39] Ali MM. Ray B. Rokeya B. Nasreen MA. Ahmed Z. Potential Healing Powers with Jute Plant- A Review. Int J of Sci: Basic and Applied Res. 2019; 48(5): 10-23.

[40] Zakaria ZA. Sulaiman MR. Gopalan HK. Ghani ZDFA. Nor RNSRM. Jais AMM. Abdullah FC. Antinociceptive and Anti-inflammatory Properties of Corchours capsularis Leaves Chloroform Extract in Experimental Animal Models. The Pharmaceutical Society of Japan. 2007; 127(2): 339-65.

[41] Quader MA. Gray AI. Waterman PG. Lavaud C. Massiot G. Hasan CM. Ahmed MD. Capsugenin 25,30-O-β-di-glucopyranoside: a new glycoside from the leaves of Corchorus capsularis. J Nat Prod. 1990; 53: 527-30.

[42] Quader MA. Ahmed M. Hasan CM. Waterman PG. Capsugenin-30-O-β-di-glucopyranoside: a new glycoside from the leaves of Corchorus capsularis. J Nat Prod. 1987; 50:479-81.

[43] Kumari N. Choudhary SB. Sharma HK. Singh BK. Kumar AA. Health-promoting properties of Corchorus leaves: A review. J Herb Med. 2019; 15: 100240.

[44] Hassan Z. Maged KG. Mekhael AG. Hanna AS. Gábor T. Helmut D. Phytochemical investigation of Corchorus olitorius and Corchorus capsularis (Family Tiliaceae) that grow in Egypt. Egyptian Pharm J. 2019; 18(2): 123-134.

[45] Jute Trading Ltd https://www.jutebag.co.uk/benefits-of-jute-leaves/. 2021.

[46] Singh P. Paul BN. Rana GC. Giri SS. Evaluation of jute leaf as feed ingredient for Labeo rohita fingerlings. Indian J Anim Nutr. 2016; 33(2): 203-207.

[47] Tareq MZ. Bashar KK. Amin MR. Sarker MDH. Moniruzzaman M. Sarker MSA. Islam MS. Nutritional composition of some jute genotypes as vegetables, International J. Vegetable Sci. 2019.

[48] Adebo HO. Ahoton LE. Quenum FJ. Adoukonou-Sagbadj H. Bello DO. Chrysostome CA. Ethnobotanical knowledge of jute (Corchorus olitorius L.) in Benin. Eur J Med Plants. 2018; 26: 1-11.

[49] Ali SI. Nasir Y. Flora of Pakistan. Karachi: Dep. of Botany, Univ. of Karachi. 1999; p. 192.

[50] Abdul S. Kaushik C. Jute-A biological elixir with multifaceted applications: An overview. Int J Res Pharm Sci. 2015; 6: 323-32.

[51] Castro MP. Obtinalla CO. Lucente ME. Evaluation of jute (Corchorus olitorius) leaves as a sushi wrap. Adv Nutr Food Sci. 2018; 3:1-5.

[52] Adekanye M. Six health benefits of jute leaf (Ewedu). https://guardian.ng/life/health-benefits-of-jute-leaf-also-known-as-ewedu-for-humans/. 2019.

[53] Zakaria ZA. Kumar GH. Nor RNS. Sulaiman MR. Fatimah CA. Mat JAM. Somchit MN. Ismail MS. Antinociceptive anti-inflammatory and antipyretic properties of an aqueous extract of Corchorus capsularis leaves in experimental animal models. Pharm Bio. 2009; 47(2): 104-110.

[54] Mondal TK. Sohel MD. Kawsar MH. Laso D. Phytochemical and antimicrobial investigations of different fractions of the methanolic extract of Corchorus capsularis leaves. African J Pharma Pharmaco. 2017; 11(17): 209-216.

[55] Nasreen MA. Ahmed Z, Ali MM. Tahmina. Determination of β-carotene in jute leaves by spectrophotometry and thin layer chromatography. World J Biol Pharm Health Sci. 2022; 9(2): 11–20.

[56] Lee YH. Choo C. Watawana MI. Jayawardena N. Waisundara VY. An appraisal of eighteen commonly consumed edible plants as functional food based on their antioxidant and starch hydrolase inhibitory activities. J Sci Food Agri. 2015; 95(14): 2956-2964.

[57] Adegoke AA. Adebayo-Tayo BC. Phytochemical composition and antimicrobial effects of Corchorous olitorius leaf extracts on four bacterial isolates. J Medicinal Plants Res. 2009; 3(3): 155-159.

[58] Ahmed Z. Ray B. Rokeya B. Ali MM. Nasreen MA. Potential Healing Powers with Jute Plant- A Review. Int J Sci: Basic Appl Res. 2019; 48: 10-23.

[59] Becer E. Soykut G. Kabadayi H. Mammadov E. ÇaliŞ İ. Vatansever S. Obtaining Stem Cell Spheroids from Foreskin Tissue and the Effect of Corchorus olitorius L. on Spheroid Proliferation. Turk J Pharm Sci. 2020; 17(3): 265-270.

[60] Tan BL. Norhaizan ME. Liew WP. Sulaiman RH. Antioxidant and Oxidative Stress: A Mutual Interplay in Age-Related Diseases. Fronti pharma. 2018; 9: 1162.

[61] USDA (US Department of Agriculture) https://fdc.nal.usda.gov/fdc-app.html#/food-details/169354/nutrients.

[62] Gomaa AA. El-Sers DA. Al-Zokeim NI. Gomaa MA. Amelioration of experimental metabolic syndrome induced in rats by orlistat and Corchorus olitorius leaf extract; role of adipo/cytokines. J Pharm Pharmacol. 2019; 71(2): 281-291.

[63] Frerejacque M. Durgeat M. Digitalis like poisons of jute seed. Compt Rend. 1954; 238: 507-509.

[64] Banergee P. Corchortoxinm a cardiac agent from jute seeds. Helv Chim Acta, 1949; 32: 2385-2392.

[65] Negm S. El-Shabrawy O. Arbid M. Radwan AS. Toxicological study of the different organs of Corchorus olitorius L. plant with special reference to their cardiac glycosides content. Zeitsc Ernaehrungsw. 1980; 19(1): 28-32.

[66] Rao EV. Rao DV. Pavanaram SK, von Euw J. Reichstein T. Structure of Corchoroside β-Glycosides and aglycones. Helv Chim Acta. 1971; 54(7): 1960-1968.

[67] Rao MR. Liu TP. Meng ZJ. Comparative studies on the cardiotonic action of corchoroside A, ouabain and strophanthin K on the heart-lung preparation of guinea pig and cat. Yao Xue Xue Bao. 1979; 14(5): 257-266.

[68] [68] Weichselbaum E. Wyness L. Stanner S. Apple polyphenols and cardiovascular disease- a review of the evidence. Nutrition Bulletin. 2010; 35(2): 92-101.

[69] Spencer JPE. The impact of fruit flavonoids on memory and cognition. British J Nutri. 2010; 104(3): S40-S47.

[70] Soykut G. Becer E. Calis I. Yucecan S. Vatansever S. Apoptotic effects of Corchorus olitorius L. leaf extracts in colon adenocarcinoma cell lines. Prog Nutri. 2018; 20(4): 689-698.

[71] Toshio F. Rong W. Katsuichiro O. Feroj AFMH. Idris MA. Akira K. Hiroshi F. Antitumor promoters in leaves of jute (Corchorus capsularis and Corchorus olitorius). Food Sci Technol Res. 2002; 8 (3): 239–243.

[72] Soykut G. Becer E. Calis I. Vatansever S. The Effect of Corchorus olitorius L. Extract on Viability of Breast Cancer Cells: A friend or a foe? Curr. Pers. MAPs. 2019; 2(2): 113-118.

[73] Valdivia HO. et al. Comparative genomics of canine-isolated Leishmania (Leishmania) amazonensis from an endemic focus of visceral leishmaniasis in Governador Valadares, southeastern. Brazil Sci Rep. 2017; 7: 40804.

[74] Pramanik PK. Chakraborti S. Bagchi A. Chakrabort T. Bioassay based Corchorus capsularis L. leaf derived β sitosterol exerts antileishmanial effects against Leishmania donovani by targeting trypanothione reductase. Scientific Reports. 2020.

[75] İşeri ÖD. Yurtcu E. Sahin FI. Haberal M. Corchorus olitorius (jute) extract induced cytotoxicity and genotoxicity on human multiple myeloma cells (ARH-77). Pharm Biol. 2013; 51(6): 766-70.

[76] Dewanjee S. Gangopadhyay M. Sahu R. Karmakar S. Cadmium induced pathophysiology: prophylactic role of edible jute (Corchorus olitorius) leaves with special emphasis on oxidative stress and mitochondrial involvement. Food Chem Toxicol. 2013; 60: 188-98.

[77] Nasreen MA. Ahmed Z. Ali MM. Shimu T. Jute leaves: A potential sources of lycopene. Int J Veg Sci. 2021; 1-9.

[78] Saliu JA. Ademiluyi AO. Boligon AA. Oboh G. Schetinger MRC. Rocha JBT. Dietary supplementation of jute leaf (Corchorus olitorius) modulates hepatic delta-aminolevulinic acid dehydratase (δ-ALAD) activity and oxidative status in high-fat fed/low streptozotocin-induced diabetic rats. J Food Biochem. 2019; 43(8): e12949.

[79] Das AK. Sahu R. Dua TK. Bag S. Gangopadhyay M. Sinha MK. Dewanjee S. Arsenic-induced myocardial injury: protective role of Corchorus olitorius leaves. Food Chem Toxicol. 2010; 48(5): 1210-7.

[80] Das AK. Bag S. Sahu R. Dua TK. Sinha MK. Gangopadhyay M. Zaman K. Dewanjee S. Protective effect of Corchorus olitorius leaves on sodium arsenite-induced toxicity in experimental rats. Food Chem Toxicol. 2010; 48(1): 326-35.

[81] Das AK. Dewanjee S. Sahu R. Dua TK. Gangopadhyay M. Sinha MK. Protective effect of Corchorus olitorius leaves against arsenic-induced oxidative stress in rat brain. Environ Toxicol Pharmacol. 2010; 29(1):64-9.

[82] Hosen SM. Das D. Kobi R. Chowdhury DU. Alam MJ. Rudra B. Bakar MA. Islam S. Rahman Z. Al-Forkan M. Study of arsenic accumulation in rice and evaluation of protective effects of Chorchorus olitorius leaves against arsenic contaminated rice induced toxicities in Wistar albino rats. BMC Pharmacol Toxicol. 2016; 17(1): 46.

[83] Yan YY. Wang YW. Chen SL. Zhuang SR. Wang CK. Anti-inflammatory effects of phenolic crude extracts from five fractions of Corchorus Olitorius L. Food Chem. 2013; 1138(2-3): 1008-14.

[84] Roy N. Role of Chorchorus capsularis phytochemicals on the feeding dynamics of Diacrisia casignetum Kollar (Lepidoptera: Arctiidae). J Entomo Zool Studies. 2014; 2(4): 227-236.

[85] Al-Snafi AE. The contents and pharmacological importance of Corchorus capsularis- A review. IOSR J Pharm. 2016; 6(6): 58-63.

[86] Oboh G. Schetinger MR. Stefanello N. Rocha JBT. Antidiabetic potentials of jute leaf (Corchorus olitorius) on type-2 diabetic rats. J Emerg Tren Engin Appl Sci. 2015; 6(7).

[87] Oboh G. Ademiluyi AO. Ayodele JA. Henle T. Jamiyu AS. Uwe S. Inhibitory effect of polyphenol-rich extracts of jute leaf (Corchorus olitorius) on key enzyme linked to type 2 diabetes (α-amylase and α-glucosidase) and hypertension (angiotensin I converting) in vitro, J Func Foods. 2012; 4(2): 450-458,

[88] Shariare MH. Noor HB. Khan JH. Uddin J. Ahamad SR. Altamimi MA. Alanazi FK. Kazi M. Liposomal drug delivery of Corchorus olitorius leaf extract containing phytol using design of experiment (DoE): In-vitro anticancer and in-vivo anti-inflammatory studies, Colloids and Surfaces B: Biointer. 2021; 199: 111543.

[89] Orieke D. Ohaeri OC. Ijeh II. Ijioma SN. Gastrointestinal and uterine smooth muscles relaxant and anti-inflammatory effects of corchorus olitorius leaf extract in laboratory animal models. J Ethnopharmacol. 2020; 247: 112224.

[90] Owoyele BV. Oyewole AL. Alimi ML. Sanni SA. Oyeleke SA. Anti-inflammatory and antipyretic properties of Corchorus olitorius aqueous root extract in Wistar rats. J Basic Clin Physiol Pharmacol. 2015; 26(4): 363-8.

[91] Tosoc JPS. Nuñeza OM. Sudha T. Darwish NHE. Mousa SA. Anticancer Effects of the Corchorus olitorius Aqueous Extract and Its Bioactive Compounds on Human Cancer Cell Lines. Molecules. 2021; 26(19): 6033.

[92] Li X. Yu Y. Xiao Y. Jin M. Hong Q. Chen A. Yang C. Protein digestibility of enzymatic hydrolysis feather meal in vitro and its application in growing pigs. Chinese J Anim. Nutr. 2012; 48 (15): 33-36.

[93] Guzzetti L. Panzeri D. Ulaszewska M. Sacco G. Forcella M. Fusi P. Tommasi N. Fiorini A. Campone L. Labra M. Assessment of Dietary Bioactive Phenolic Compounds and Agricultural Sustainability of an African Leafy Vegetable Corchorus olitorius L. Front Nutr. 2021; 8: 667812.

[94] Hamid K. Saha MR. Urmi KF. Habib MR. Rahman MM. Screening of different parts of the plant Pandanus odorus for its antioxidant activity. Int J Appl Biol Pharm. 2010; 1:1364–8.

[95] Steinmetz KA. Potter JD. Vegetables, fruit, and cancer prevention: a review. J Am Diet Assoc. 1996; 96: 1027–39.

[96] Oboh G. Raddatz H. Henle T. Characterization of the antioxidant properties of hydrophilic and lipophilic extracts of Jute (Corchorus olitorius) leaf. Int J Food Sci Nutr. 60 Suppl. 2009; 2:124-34.

[97] Azeez L. Adeoye MD. Majolagbe TA. Lawal AT. Badiru R. Antioxidant activity and phytochemical contents of some selected Nigerian fruits and vegetables. American J Chem. 2012; 2(4): 209-213.

[98] Henriques F. Guiné R. Barroca MJ. Chemical properties of pumpkin dried by different methods. Crotian J Food Tech Biotech Nutri. 2012; 7(1-2): 98-105.

[99] Bennett LE. Jegasothy H. Konczak I. Frank D. Sudharmarajan S. Clingeleffer PR. Total polyphenolics and anti-oxidant properties of selected dried fruits and relationships to drying conditions. J Func Foods. 2011; 3: 115-124.

[100] Nath R. Roy S. De B. Duttachoudhury M. Anticancer and antioxidant activity of Croton: a review. Inter J Pharma Pharma Sci. 2013; 5: 63-70.

[101] Sun J. Chu YF. Wu X. Liu RH. Antioxidant and antiproliferative activities of common fruits. J Agri Food Chem. 2002; 50(25): 7449-7454.

[102] Gharras HE. Polyphenols: food sources, properties and applications-a review. Int J Food Sci Tech. 2009; 44(12): 2512-2518.

[103] Zhou Y. Jiang Z. Lu H. Xu Z. Tong R. Shi J. Jia G. Recent Advances of Natural Polyphenols Activators for Keap1-Nrf2 Signaling Pathway. Chem Biodiver. 2019; 16(11): e1900400.

[104] Hider RC. Liu ZD. Khodr HH. Metal chelation of polyphenols. Methods in Enzymo. 2001; 335: 190-203.

[105] Scalbert A. Johnson IT. Saltmarsh M. Polyphenols: antioxidants and beyond. American J Clin Nutri. 2005; 81(1): 215S-217S.

[106] Pandey KB. Rizvi SI. Plant polyphenols as dietary antioxidants in human health and disease. Oxidative Med Cellul Longevity. 2009; 2(5): 270-278.

[107] Omenna EC. Ojo AI. Comparative phytochemical screening of kenaf and jute leaves. J Nutr Health Food Eng. 2018; 8(5): 366–369.

[108] Alexopoulou E. Papattheohari Y. Christou M. et al. Kenaf: A multi–purpose crop for sever industrial applications. London, Springer–verlag. 2013; p. 1–15.

[109] Jaihyunk R. Soon–Jae K. Joon–Woo A. et al. Phytochemicals and antioxidant activity in the kenaf plant (Hibiscus cannabinus L.). J Plant Biotech. 2017; 44:191–202.

[110] İşeri ÖD. Körpe DA. Sahin FI. Haberal M. Corchorus olitorius and Urtica pilulifera extracts alleviate copper induced oxidative damage and genotoxicity in tomato. Acta Biol Hung. 2018; 69(3): 300-312.

[111] Houghton PJ. Hylands PJ. Mensah AY. Hensel A. Deters AM. In vitro tests and ethnopharmacological investigations: Wound healing as an example. J Ethnopharmacol. 2005; 100: 100-107.

[112] Hamid AA. Aiyelaagbe OO. Usman LA. Ameen OM. Lawa A. Antioxidants: Its medicinal and pharmacological applications. African J Pure Appl Chem. 2010; 4(8): 142-151.

[113] Vya B. Boye A. Quansah N. Antioxidant and wound healing studies on the extracts of Corchorus olitorius leaf. Sci Res Review J. 2013; 1(3): 67-73.

[114] Sadat A. Hore M. Chakraborty K. Roy S. Phytochemical analysis and antioxidant activity of methanolic extract of leaves of Corchorus olitorius. Int J Curr Pharm Res. 2017; 9(5): 59-63.

[115] [115] Lorenzen S. Thuss-Patience P. Al-Batran SE. et al. Impact of Pathologic Complete Response on Disease-Free Survival in Patients with Esophagogastric Adenocarcinoma

Receiving Preoperative Docetaxel-Based Chemotherapy. Annals of Oncology, 2013; 24: 2068-2073.

[116] Mahmoud AS. Thao N. Mario A. Corchorus olitorius Linn: A Rich Source of Ω3-Fatty Acids. Pharm Anal Acta. 2016; 7(6).

[117] Wagdy R. Abdelkader RM. El-Khatib AH. Linscheid MW. Hamdi N. Handoussa H. Neuromodulatory Activity of Dietary Phenolics Derived from Corchorus olitorius L. J Food Sci. 2019; 84(5): 1012-1022.

[118] Özlem Dİ. Erkan Y. Feride IS. Mehmet H. Corchorus olitorius (jute) extract induced cytotoxicity and genotoxicity on human multiple myeloma cells (ARH-77). Pharma Biol, 2013; 51: 6, 766-770,

[119] Senu AJ. Olusola EM. Afolashade OO. The In-vitro effect of Corchorus olitorius (Linn.) on the antibacterial activities of five antibiotics. Global Res J Microbiol. 2012; 2(2): 113.

[120] Ahmed Z. Akhter F. Alamgir M. Isolation of some jute retting fungi from natural source and their effect on dry ribbon retting of CVL-1 jute (Chorchorus capsularis), Bangladesh J Sci Res. 1999; 17: 107-109.

[121] Yakoub ARB. Abdehedi O. Jridi M. Elfalleh W. Nasri M. Ferchichi A. Flavonoids, phenols, antioxidant, and antimicrobial activities in various extracts from Tossa jute leave (Corchorus olitorus L.). Indus Crops Products. 2018; 118: 206–213.

[122] Hayder TH. Enas JK. Phytochemical investigation of Corchorus olitorius L. leaves cultivated in Iraq and it's In vitro antiviral activity. Iraqi J Pharm Sci. 2018; 27(2): 115-122.

[123] Talukder D. Khanam LAM. Toxicity of four plant based products against three stored product pests. J Bio Sci. 2009; 17: 149-153.

[124] Duke JA. Wain KK. Medicinal plants of the world. Computer index with more than 85,000 entries. 1981; 3 vols.

[125] Sawle BD. Salimath B. Deshpande R. Bedre MD. Prabhakar BK. Venkataraman A. Biosynthesis and stabilization of Au and Au–Ag alloy nanoparticles by fungus, Fusarium semitectum. Sci Technol Adv Mater. 2008; 9: 035012.

[126] Hussain I. Singh NB. Singh A. Singh H. Singh SC. Green synthesis of nanoparticles and its potential application. Biotech Lett. 2016; 38: 545–560.

[127] Ismail EH. Saqer AMA. Assirey E. Naqvi A. Okasha RM. Successful Green Synthesis of Gold Nanoparticles using a Corchorus olitorius Extract and Their Antiproliferative Effect in Cancer Cells. Int J Mol Sci. 2018; 19(9): 2612.

[128] Kunoh T. Takeda M. Matsumoto S. Suzuki I. Takano M. Kunoh H. Takada J. Green Synthesis of Gold Nanoparticles Coupled with Nucleic Acid Oxidation. ACS Sustain. Chem Eng 2018; 6: 364–373.

[129] Shankar SS. Rai A. Ahmad A. Sastry M. Rapid synthesis of Au, Ag, and bimetallic Au core-Ag shell nanoparticles using Neem (Azadirachta indica) leaf broth. J Colloid Interface Sci. 2004; 275: 496–502.

[130] Philip D. Green synthesis of gold and silver nanoparticles using Hibiscus rosa sinensis. Phys E Low Dimens Syst Nanostruct. 2010; 42: 1417-1424.

[131] Fayaz AM. Girilal M. Venkatesan R. Kalaichelvan PT Biosynthesis of anisotropic gold nanoparticles using Maduca longifolia extract and their potential in infrared absorption. Colloids Surf B Biointer. 2011; 88: 287–291.

[132] Jayaseelan C. Ramkumar R. Rahuman AA. Perumal P. Green synthesis of gold nanoparticles using seed aqueous extract of Abelmoschus esculentus and its antifungal activity. Ind Crops Prod. 2013; 45: 423–429.

[133] Gopinath K. Venkatesh K.S. Ilangovan R. Sankaranarayanan K. Arumugam A Green synthesis of gold nanoparticles from leaf extract of Terminalia arjuna for the enhanced mitotic cell division and pollen germination activity. Ind Crops Prod. 2013; 50: 737–742.

[134] Ganeshkumar M. Sathishkumar M. Ponrasu T. Dinesh MG. Suguna L. Spontaneous ultra fast synthesis of gold nanoparticles using Punica granatum for cancer targeted drug delivery. Colloids Surf B Biointer. 2013; 106: 208–216.

[135] Nahrstedt A. Butterweck V. Lessons learned from herbal medicinal products: The example of St. John's Wort. J Nat Prod. 2010; 73: 1015–1021.

[136] Qiao X. Ye M. Xiang C. Wang Q. Liu CF. Miao WJ. Guo DA. Analytical strategy to reveal the in vivo process of multi-component herbal medicine: A pharmacokinetic study of licorice using liquid chromatography coupled with triple quadrupole mass spectrometry. J Chromato. A. 2012; 1258: 84–93.

[137] Esimone CO. Nwafor SV. Okoli CO. Chah KF. Uzuegbu DB. Chibundu C. Eche MA. Adikwu MU. In vivo evaluation of interaction between aqueous seed extract of Garcinia kola Heckel and ciprofloxacin hydrochloride. Am J Ther. 2002; 9: 275–280.

[138] Gunaratna PC. Kissinger PT. Kissinger CB. Gitzen JF. An automated blood sampler for simultaneous sampling of systemic blood and brain microdialysates for drug absorption, distribution, metabolism, and elimination studies. J Pharma Toxicol Methods. 2004; 49: 57–64.

[139] Anwer MM. Sarkar S. Alam MM. Hossain SMM. Hoque MM. Haque MT. Saifullah K. Influence of Low Temperature Argon and Oxygen Plasma Treatment on the Band-gap of Jute. Int J Engi Appl Sci. 2018; 5(3): 79-82.

[140] Anwer MM. Kayser MA. Sarkar S. Hossain SMM. Akter S. Kamal TB. Jahan FN. Fourier Transform Infrared Spectroscopic Analyses of Argon and Oxygen Plasma Treated Jute. Int J Engi Appl Sci. 2018; 5(7): 5-8.

[141] Anwer MM. Kayser MA. Sarkar S. Hossain SMM. Akter S. Kamal TB. Jahan FN. Scanning Electron Microscopy Analysis of Argon Plasma Treated Jute Fiber. Int J Engi Appl Sci. 2018; 5(7): 9-11.

[142] Mikhailova EO. Gold Nanoparticles: Biosynthesis and Potential of Biomedical Application. J Fun Biomat. 2021; 12(4): 70.

2.2.4.2.1 Lycopene from jute leaf [Aleya et al 2021, Aleya et al 2022]

2.2.4.2.1.1 Introduction

Lycopene is a phytochemical, implicated in protecting humans from damage due to free radicals (**Komuroglu et al 2018; Mascio et al 1989; Rao and Ali2007**). Lycopene is associated with inhibition of disease in humans (**Arab and Steck 2000; Fuhrman et al 1997; Mendelová et al2013; Palozza et al2012; Rissanen et al 2003**) by protecting lipids, low-density lipoproteins and DNA from oxidative damage (**Ansari and Gupta 2003; Dizdaroglu and Jaruga 2012; Giovannucci 2002**).Lycopene content research (**Fish et al 2002**) has focused on tomato (*Solanum lycopersicum*) and some red color fruit including watermelon [*Citrullus lanatus* (Thunb.) Matsum & Nakai]. Tomato and tomato-based products are considered the main source of lycopene and more than 80% of dietary lycopene is derived from processed tomato products (**Storniolo et al2019**). Watermelon is another vegetable dietary source of lycopene (**Fish and Davis 2003; Suwanaruang 2016**). Some non-red and non-orange plants such as asparagus (*Asparagus officinalis*) and parsley (Petroselinum crispum Mill.) are also good dietary sources of lycopene (**Hedayati et al 2019**). Humans cannot synthesize lycopene (**Kirpichenkova et al 2018**), it must be consumed in a daily diet (**Woodside et al 2015**).Young leaves of cultivated jute species, *Corchorus olitorius* and *C. capsularis* are consumed fresh or processed (**Furumuto et al 2002; Li et al 2012**). Little information is available regarding lycopene in leafy vege-tables; spinach (*Spinacia oleracea*) and lettuce (*Lactuca sativa*) were examined but no lycopene was found (**Brown et al 2004**). Lycopene content in jute leaves has not been investigated. Work was undertaken to determine lycopene content in fresh jute leaves of different ages and how lycopene content was affected by processing and preservation.

2.2.4.2.1.2 Materials and methods

The experiment was undertaken at the Jute Agricultural Experimental Station in Manikganj, Bangladesh Jute Research Institute, Dhaka, Bangladesh. The silt loam soil was prepared by 4–5 plowings followed by laddering. Raw seeds of *C. olitorius*, cv. O-9897, and *C. capsularis*, cv. CVL-1 were sown in 3 m length lines with line-to-line distance of 30 cm, and plant-to-plant distance of 8–10cm. Cultural activities were according to Chowdhury and Hassan (**2013**).Fresh jute leaves were harvested from 45, 60 and 120 days old plants for estimation of lycopene content and the experiments were terminated. A single line was harvested at each date of leaves collection. To compare the lycopene content of jute leaves with tomato (var. BARI Tomato 7), matured green tomato (before ripening) and red tomato (ripen) were collected from a local market. Three replications were used for each sample. Other jute leaves werecollected at 20 days from a single line of plants from the same experimental field. Two g of 20-day-old fresh jute leaves were processed by-

(i) boiling in 10 mL distilled water for 15 min;

(ii) frying with 5 mL soybean oil for 15min;

(iii) boiling in 10mL distilled water with 0.1g salt for 15min;

(iv) frying with 5mL soybean oil with 0.1g salt for 15min,

(v) drying in a convection oven (Memmert GmbH, Schwabach, Germany) at 60°C for 3h, and (vi) Soaking in 2.5mL of vinegar for 5min. Fresh leaves without processing were controls.Leaves from 120-day-old plants were used for preservation.

The drying methods were as follows-

i) drying in a convection oven (Memmert-GmbH) at 100°C overnight,

ii) cold drying leaves were kept under refrigeration at 4°C in brown paper envelopes for ≥7 days so leaves can dry gradually, and

iii) air drying leaves were left in the open at room temperature (33–35°C).

Dried leaves were ground with a mortar and pestle, kept in plastic bottles and placed in a freezer (−20°C). Lycopene content was estimated after 3 years of preservation. The experiment was started at the same time and the data for the non-preserved leaves withheld for 3 years.Lycopene content in jute leaves samples was extracted with a hexane:ethanol:acetone (2:1:1)($^{v}/_{v}$) mix (**Sharma and Le Maguer 1996**). Two-g of leaves were chopped and added to 50 mL of solvent [hexane:ethanol:acetone (2:1:1) ($^{v}/_{v}$)] in a 250mL Erlenmeyer flask. Flasks were placed on a shaker for 30 min at 100 rpm. Then, 20 mL of deionized water was added to the solution and shaking continued for 2 min. The mix was left to separate into aqueous and organic layers. The organic layer was used for lycopene estimation. A similar procedure wasfollowed for the estimation of lycopene content in fresh green and red tomatoes.

Since lycopene exists in different sizes, absorbance was measured at 472nm (for comparatively shorter chain molecules of lycopene) and 502nm (for longer chain molecules of lycopene) with a spectrophotometer using hexane as a blank. The lycopene concentration was calculated using its specific extinction coefficient (E 1%, 1cm) of 3450 in hexane at 472nm and 3150 at 502nm. For all determinations, there were three biological replications and three technical replications (repeated measurements of the same sample). The lycopene content in the sample was determined using the formulae of Alda et al (**2009**).Data were subjected to analysis of variance using MSTAT-C (ver. 6.0,Michigan State University, East Lansing, MI). If interactions were significant, they were used to explain the results. If interactions were not significant, means were separated using least significant differences.

2.2.4.2.1.3 Results[Aleya et al 2021, Aleya et al 2022]

Species (S), plant age (A), nanometer (N), and all interaction affected lycopene content (**Table 2.2.4.2.1.1**). Lycopene content in fresh jute leaves varied with plant age. The highest lycopene content was from 45-day-old plant leaves and it gradually declined in older plant leaves in both species (**Table 2.2.4.2.1.2**). The lowest lycopene content was in 120-day-old plant leaves in both species.

Table 2.2.4.2.1.1 Combined analysis of variance of leaf lycopene content for two cultivated jute species at three plant ages with values estimated at 472 and 502 nm absorbance.[Aleya et al 2021, Aleya et al 2022]

Source	Degrees of freedom	Sum of squares	Mean of squares
Species (S)	1	301.312	301.312**
Plant ages (A)	2	4171.297	2085.649**
S × A	2	69.064	34.532**
Nanometer (N)	1	3975.933	3975.933**
S × N	1	25.184	25.184**
A × N	2	712.045	356.023**
S × A × N	2	96.295	48.148**
Error	22	6.578	0.299
Total	35	9357.992	267.371

**Significance at $P < 0.01$.

Table 2.2.4.2.1.2. Interaction[a] response for leaf lycopene content of cultivated jute species at three plant ages with values estimated at 472 and 502 nm absorbance.[Aleya et al 2021, Aleya et al 2022]

Species	Absorbance (nm)	Plant age (days)	Lycopene content (mg/100 g)
Corchorus capsularis	472	45	50.99b[b]
		60	47.26 c
		120	13.66 g
	502	45	26.18e
		60	11.80 h
		120	5.85i
Corchorus olitorius	472	45	53.55a
		60	47.41 c
		120	23.29 f
	502	45	27.85d
		60	24.54 f
		120	13.82 g

[a]Data in the interaction analyzed with Least Squares Means and mean within a column for lycopene content separated with LSD.

[b]Values in the column followed by the same letter are not significantly different, $P < 0.01$.

Table 2.2.4.2.1.3. Lycopene content of red and green tomatoes evaluated at 472 and 502 nm absorbance.[Aleya et al 2021, Aleya et al 2022]

	Lycopene (mg/100 g)	
Tomato color	472 nm	502 nm
Red ripe	15.50 ± 0.14^a	12.29 ± 0.12^a
Green	4.43 ± 0.12^b	1.43 ± 0.09^b

[a]values within a column followed by the same letter are not significantly different at $P < 0.01$ according to LSD.

Red tomato had more lycopene than green tomato (**Table 2.2.4.2.1.3**). Leaves of 45 days old *C. olitorius* plant contain more than double the lycopene than red tomato, but amounts of lycopene in tomatoes were similar to jute leaves at 120 days (**Fig2.2.4.2.1.1**).

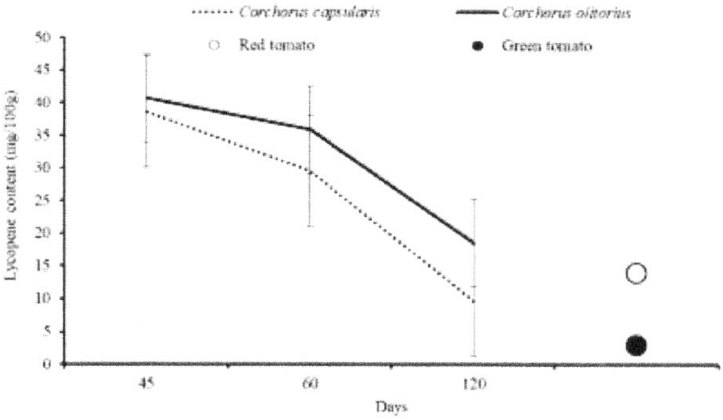

Figure 2.2.4.2.1.1. Lycopene content of jute leaves of 2 cultivated species at 3 plant ages compared to red and green tomatoes. [Aleya et al 2021, Aleya et al 2022]

Table 2.2.4.2.1.4. Combined analysis of variance of leaf lycopene content for cultivatedjute species for processed leaves with values estimated at 472 and 502 nmabsorbance.[Aleya et al 2021, Aleya et al 2022]

Source	df	Sum of squares	Mean of Squares
Species (S)	1	157.577	157.577**
Processed leaves (P)	6	145.502	24.250**
S × P	6	13.357	2.226**
Nanometer (N)	1	879.144	879.144**
S × N	1	21.716	21.716**
P × N	6	115.669	19.278**
S × P × N	6	15.095	2.516**
Error	54	1.334	0.025
Total	83	1349.446	3.024

**Significant at $P < 0.01$.

Table 2.2.4.2.1.5. Interaction[a] effect on leaf lycopene content of cultivated jute species subjected to six processing treatments with values estimated at 472 and 502nm absorbance.[Aleya et al 2021, Aleya et al 2022]

Species	Absorbance (nm)	Processing method	Lycopene content (mg/100 g)
C. capsularis	472	Boil with DH_2O for 15 min	6.95h[b]
		Fry with soybean oil for 15 min	4.40k
		Boil with DH_2O + salt for 15 min	12.16 c
		Fry with soybean oil + salt for 15 min	6.09 j
		Dried at 60°C for 3hrs in oven	6.25 j
		Soaked in vinegar for 5 min	6.60i
		Fresh leaves as control	3.74 l
	502	Boil with DH_2O for 15 min	0.63q
		Fry with soybean oil for 15 min	0.50q
		Boil with DH_2O + salt for 15 min	0.57q
		Fry with soybean oil + salt for 15 min	2.86 m
		Dried at 60°C for 3hrs in oven	1.55o
		Soaked in vinegar for 5 min	1.15p
		Fresh leaves as control	0.76q
C. olitorius	472	Boil with DH_2O for 15 min	11.09d
		Fry with soybean oil for 15 min	7.82 g
		Boil with DH_2O + salt for 15 min	14.77b
		Fry with soybean oil + salt for 15 min	9.18 f
		Dried at 60°C for 3hrs in oven	12.53b
		Soaked in vinegar for 5 min	21.15a
		Fresh leaves as control	6.74hi
	502	Boil with DH_2O for 15 min	3.66 l
		Fry with soybean oil for 15 min	3.01 l
		Boil with DH_2O + salt for 15 min	3.52 l
		Fry with soybean oil + salt for 15 min	2.20 n
		Dried at 60°C for 3hrs in oven	2.75 m
		Soaked in vinegar for 5 min	6.09 j
		Fresh leaves as control	2.87 m

[a]Data in the interaction analyzed with Least Squares Means and mean within a column for lycopene content separated with LSD.

[b]Values in the column followed by the same letter are not significantly different, $P < 0.01$.

Species (S), process technique (P), nanometer (N) and all interactions affected lycopene content (**Table 2.2.4.2.1.4**). Lycopene content varied during processes (**Table 2.2.4.2.1.5**). There was an increase in lycopene content in processed leaves compared to fresh leaves (control).

The trend varied with processing method. Most of the lycopene came from vinegar-soaked jute leaves of *C. olitorius*, followed by boiling with salt in *C. olitorius* leaves. The addition of salt during boiling increased the availability of lycopene than boiling without salt at 472nm in both species. A slight decrease occurred in the lycopene content in salt added, boiled leaves, than without salt and boiled, when measured at 502nm in both species. A similar trend occurred in oil fried leaves, and addition of salt with oil fried leaves, in both species. The lowest lycopene content was in only oil fried leaves. Oven dried jute leaves, as a processing method, had lower lycopene content in both species than vinegar soaked leaves and boiling in water and salt added leaves.Drying is a simple technique for preservation of jute leaves to ensure availability during the off-season. Species (S), preservation (D), nanometer (N), and all interactions affected lycopene content (**Table 2.2.4.2.1.6**). Among the three drying processes, cold dried preserved leaves had more lycopene (**Table 2.2.4.2.1.7**).

Table 2.2.4.2.1.6. Combined analysis of variance of leaf lycopene content for cultivated jute species subjected to three preservation methods and estimated at 472 and 502nm absorbance. [Aleya et al 2021, Aleya et al 2022]

Source	Degrees of freedom	Sum of squares	Mean of Squares
Species (S)	1	391.519	391.519**
Preserved leaves (D)	2	1132.454	566.227**
S × D	2	28.447	14.223**
Nanometer (N)	1	2736.667	2736.667**
S × N	1	113.277	113.277**
D × N	2	701.439	350.719**
S × D × N	2	22.361	11.181**
Error	22	0.807	0.037
Total	35	5126.972	146.485

** significant at $P < 0.01$.

Table 2.2.4.2.1.7. Interaction[a] results for lycopene content of leaves of cultivated jute species subjected to 3 preservation methods estimated at 472 and 502 nm absorbance. [Aleya et al 2021, Aleya et al 2022]

Species	Absorbance (nm)	Preservation method	Lycopene content (mg/100 g)
C. capsularis	472	Oven dried	14.23e[b]
		Air dried	15.02d
		Cold dried	36.11b
	502	Oven dried	5.82k
		Air dried	6.39 j
		Cold dried	11.48 g
C. olitorius	472	Oven dried	24.81 c
		Air dried	25.23 c
		Cold dried	45.75a
	502	Oven dried	8.42i
		Air dried	13.73 f
		Cold dried	15.68d

[a]Data in the interaction analyzed with Least Squares Means and mean within a column for lycopene content separated with LSD.

[b]Values in the column followed by the same letter are not significantly different, $P < 0.01$.

2.2.4.2.1.4 Discussion [Aleya et al 2021, Aleya et al 2022]

Lycopene content decreased in jute leaves due to increasing plant age. Forty- five-day-old plant leaves had more lycopene than older plant leaves. The changes in lycopene content at different leaf ages may be due to its capacity to utilize carbon during photosynthesis. Younger plants leaves utilize more carbon for synthesizing lycopene and older plant leaves have less capacity to store carbon, which may hamper lycopene synthesis (**Dhami et al 2018**). Little has been reported on lycopene content in plant leaves. We observed that lycopene content was lower in green compared to red tomatoes. Lycopene content in fruit is influenced by degree of ripeness (**Rocha et al 2015; Sgherri et al 2015**) and total chlorophyll levels decrease during fruit ripening, whereas lycopene levels increase (**Sandei 2018**).

The processing method affected the availability of lycopene compared to fresh leaves. Variation of lycopene bioavailability between species may be due to genetic make-up of the species. These two species originated from different regions, *C. olitorius* from Africa and *C. capsularis* from Indo-Burma, including South China. They are distinct in a number of agronomic and biomass- associated traits as well as fiber and seed coat color (**Islam et al**

2017). Lycopene bioavailability increased in processed jute leaves compared to fresh leaves because most processing methods included heat which improves bioavailability of lycopene. Bioavailability of lycopene is higher in tomato paste or puree (produced throughheat) compared to fresh tomato (**Gärtner et al 1997; Porrini et al 1998**). Lycopene content is influenced by storage and processing method (**Rocha et al 2015; Sgherri et al 2015**). Differences occurred in lycopene content among drying processes of the 3 year preserved leaves (**Table 2.2.4.2.1.7**). Cold drying leaves retain the highest lycopene since freezing can remove water with-out disruption of cells and retain the nutritional quality (**Marques et al 2007**).This is likely the first report of lycopene content estimation of fresh jute leaves as well as the effects of processing and preservation of lycopene content in jute leaves. Regular consumption of jute leaves could provide lycopene to the human diet. However, it is necessary to determine if consumption and digestion affect the amount available for human diet.

2.2.4.2.1.5 References

Alda, L.M., I. Gogoaşă, D.M. Bordean, I. Gergen, S. Alda, C. Moldovan, and L. Niţă. 2009. Lycopene content of tomatoes and tomato products. Journal of Agroalimentary Processes and Technologies 15(4):540–542.

Ansari, M.S., and N.P. Gupta. 2003. A comparison of lycopene and orchidectomy vs orchidectomy alone in the management of advanced prostate cancer. British Journal of Urology International 92(4):375–378. doi: 10.1046/j.1464-410X.2003.04370.x.

Arab, L., and S. Steck. 2000. Lycopene and cardiovascular disease. Am. J. Clin. Nutr. 71 (6):1691S–1695S. doi: 10.1093/ajcn/71.6.1691S.

Brown, M.J., M.G. Ferruzzi, M.L. Nguyen, D.A. Cooper, A.L. Eldridge, S.J. Schwartz, and W.

S. White. 2004. Carotenoid bioavailability is higher from salads ingested with full-fat than with fat-reduced salad dressings as measured with electrochemical detection. American Journal of Clinical Nutrition 80(2):396–403. doi: 10.1093/ajcn/80.2.396.

Chowdhury, M.A.H., and M.S. Hassan. 2013. Hand book of agricultural technology. Bangladesh Agricultural Research Council, Dhaka, Bangladesh.

Dhami, N., T.T. David, and I.C. Christopher. 2018. Leaf-age dependent response of carotenoid accumulation to elevated CO 2 in Arabidopsis. Arch. Biochem. Biophys. 647:67–75. doi: 10.1016/j.abb.2018.03.034.

Dizdaroglu, M., and P. Jaruga. 2012. Mechanisms of free radical-induced damage to DNA. Free Radic. Res. 46(4):382–419. doi: 10.3109/10715762.2011.653969.

Fish, W.W., and A.R. Davis. 2003. The effects of frozen storage conditions on lycopene stability in watermelon tissue. J. Agric. Food Chem. 51(12):3582–3585. doi: 10.1021/jf030022f.

Fish, W.W., P. Perkins-Veazie, and J.K. Collins. 2002. A quantitative assay for lycopene that utilizes reduced volumes of organic solvents. Journal of Food Composition and Analysis 15 (3):309–317. doi: 10.1006/jfca.2002.1069.

Fuhrman, B., A. Elis, and M. Aviram. 1997. Hypocholesterolemic effect of lycopene and β-carotene is related to suppression of cholesterol synthesis and augmentation of LDL receptor activity in macrophages. Biochem. Biophys. Res. Commun. 233(3):658–662. doi: 10.1006/ bbrc.1997.6520.

Furumuto, T., R. Wang, K. Okazaki, F.A. Hasan, I.M. Ali, A. Kondo, and H. Fukui. 2002. Antitumor promoters in leaves of jute (Corchorus capsularisand Corchorus olitorius). Food Science and Technology Research 8(3):239–243. doi: 10.3136/fstr.8.239.

Gärtner, C., W. Stahl, and H. Sies. 1997. Lycopene is more bioavailable from tomato paste than from fresh tomatoes. American Journal of Clinical Nutrition 66(1):116–122. doi: 10.1093/ ajcn/66.1.116.

Giovannucci, E. 2002. A review of epidemiologic studies of tomatoes. Lycopene, and Prostate Cancer. Experimental Biology and Medicine 227(10):852–859.

Hedayati, N., M.B. Naeini, A. Nezami, H. Hosseinzadeh, A.W. Hayes, S. Hosseini, M. Imenshahidi, and G. Karimi. 2019. Protective effect of lycopene against chemical and natural toxins: A review. Bio Factors 45:5–23.

Islam, M.S., J.A. Saito, E.M. Emdad, B. Ahmed, M.M. Islam, A. Halim, Q.M.M. Hossen, M. Z. Hossain, R. Ahmed, M.S. Hossain, et al. 2017. Comparative genomics of two jute species and insight into fibre biogenesis. Nature Plants. 32. Articlenumber 16223. doi: 10.1038/ nplants.2016. 223

Kirpichenkova, E.V., A.A. Korolev, G.G. Onishchenko, E.I. Nikitenko, E.L. Denisova, R. N. Fetisov, E.S. Petrove, and E.A. Fanda. 2018. Study of consumption frequency of the main source of lycopene and its quantification in students' diet. I.P. Pavlov Russian Biological Herald 26(4):474–483. doi: 10.23888/PAVLOVJ2018264474-483.

Komuroglu, A.U., F. Yur, and I.H. Ekin. 2018. Influence of lycopene administration on neopterin. Myeloperoxidase and Gamma Glutamyl Transferase in Diabetic Rats. Fresenius Environmental Bulletin 27(6):4285–4292.

Li, C.J., S.Y. Huang, M.Y. Wu, Y.C. Chen, S.F. Tsang, J.H. Chyuan, and H.Y. Hsu. 2012. Induction of apoptosis by ethanolic extract of Corchorus olitorius leaf in human hepatocellular carcinoma (HepG2) cells via a mitochondria-dependent pathway. Molecules 17(8):9348–9360. doi: 10.3390/molecules17089348.

Marques, L.G., M.C. Ferreira, and J.T. Freire. 2007. Freeze-drying of acerola (Malpighia glabra L.). Chemical Engineering and Processing: Process Intensification 46(5):451–457. doi: 10.1016/j.cep.2006.04.011.

Mascio, P.D., S. Kaiser, and H. Sies. 1989. Lycopene as the most efficient biological carotenoid singlet oxygen quencher. Arch. Biochem. Biophys. 274(2):532–538. doi: 10.1016/0003- 9861(89)90467-0.

Mendelová, A., M. Fikselová, and Ľ. Mendel. 2013. Carotenoids and lycopene content in fresh and dried tomato fruits and tomato juice. Acta Universitatis Agriculturaeet Silviculturae Mendelianae Brunensis 61(5):1329–1337. doi: 10.11118/actaun201361051329.

Palozza, P., A. Catalano, R.E. Simone, M.C. Mele, and A. Cittadini. 2012. Effect of lycopene and tomato products on cholesterol metabolism. Ann. Nutr. Metab. 61(2):126–134. doi: 10.1159/ 000342077.

Porrini, M., P. Riso, and G. Testolin. 1998. Absorption of lycopene from single or daily portions of raw and processed tomato. British Journal of Nutrition 80(4):353–361. doi: 10.1017/S000711459800141X.

Rao, A.V., and A. Ali. 2007. Biologically active phytochemicals in human health: Lycopene. International Journal of Food Properties 10(2):279–288. doi: 10.1080/10942910601052673.

Rissanen, T.H., S. Voutilainen, K. Nyyssönen, R. Salonen, G.A. Kaplan, and J.T. Salonen. 2003. Serum lycopene concentrations and carotid atherosclerosis: The Kuopio Ischaemic heart disease risk factor study. American Journal of Clinical Nutrition 77(1):133–138. doi: 10.1093/ajcn/77.1.133.

Rocha, C.E., R. Roehrs, and M. Roehrs. 2015. Chemistry, sources and benefits of lycopene, p. 105–114. In: J.R. Bailey (ed.). Lycopene: Food sources, potential role in human health and antioxidant effects. Nova Science Inc., New York.

Sandei, L. 2018. Lycopene and tomatoes, p. 149–178. In: A. Venketeshwer Rao, G.L. Young, and L.G. Rao (eds.). Lycopene and tomatoes in human nutrition and health. CRC Press, New York.

Sgherri, C., U. Pérez-López, and C. Pinzino. 2015. Antioxidant properties of food products containing lycopene are increased by the presence of chlorophyll, p. 39–90. In: J.R. Bailey (ed.). Lycopene: Food sources, potential role in human health and antioxidant effects. Nova Science Inc, New York.

Sharma, S.K., and M. Le Maguer. 1996. Lycopene in tomatoes and tomato pulp fractions. Italian Journal of Food Science 8(2):107–113.

Storniolo, C.E., I. Sacanella, M.T. Mitjavila, R.M. Lamuela-Raventos, and J.J. Moreno. 2019. Bioactive compounds of cooked tomato sauce modulate oxidative stress and arachidonic acid cascade induced by oxidized LDL in macrophage cultures. Nutrients. 11(8):1880. ArticleID number 1880 doi:10.3390/nu11081880.

Suwanaruang, T. 2016. Analyzing lycopene content in fruits. Agriculture and Agricultural Science Procedia. 11:46–48. doi: 10.1016/j.aaspro.2016.12.008.

Woodside, J.V., A.J. McGrath, N. Lyner, and M.C. McKinley. 2015. Carotenoids and health in older people. Maturitas 80(1):63–68. doi: 10.1016/j.maturitas.2014.10.012.

2.2.4.2.2L-ascorbic acidfrom jute leaf [Aleya et al 2021, Aleya et al 2022]

2.2.4.2.2.1 Introduction

Jute (*Corchorus* sp.) leaves are consumed as vegetables in many tropical countries. It has more than 100 species [**1**], among them only two species namely, *Corchorus olitorius* and *Corchorus capsularis* are commercially cultivated for fibre production. Along with fibre production, jute is cultivated as vegetables which commonly called 'pat shak' in Bangladesh. Jute leaves are rich source of vitamins, antioxidants and phenolic compounds [**2, 3**]. It contains L-ascorbic acid [**4, 5**] as well. L-ascorbic acid is known as Vitamin-C, found in variable quantities in fruits and vegetables [**6**]. L-ascorbic acid or vitamin C provide numerous health benefit such as prevention of scurvy, functions in collagen, reduction in plasma cholesterol level, enhancement of immune system and reaction with singlet oxygen and other free radicals. As an important antioxidant, L-ascorbic reduces risk of cardiovascular diseases and some form of cancers [**7, 8**].The human body cannot synthesize vitamin C endogenously, and it must intake through dietary component [**9**]. Like other leafy vegetables jute leaves are to cook. During cooking ascorbic acids are destroyed at high temperature, since it easily leaches into cooking water being a water-soluble vitamin. Losses of ascorbic acid depend upon cooking methods and periods [**10**]. The actual amount of vitamin C exists in the fresh condition is reduced through cooking. Therefore, it is important to investigate L-ascorbic status in cooked jute leaves. Jute leaves are dried and preserved for off season use. Drying is the technique of preservation for off season consumption as well as for year-round availabilityto commercialization. Drying and cooking process affect the L-ascorbic content. Therefore, this experiment has been carried out to investigate the L-ascorbic acid status in i) fresh jute leaves of different varieties of Corchorus olitorius and Corchorus capsularis ii) cooked and processed jute leaves and iii) preserved jute leaves.

2.2.4.2.2.2 Material and methods [Aleya et al 2021, Aleya et al 2022]

Jute leaves (*C. olitorius* and *C. capsularis*) of different varieties were collected from the experimental field of Breeding Division of Bangladesh Jute Research Institute, Dhaka, Bangladesh. The site belongs to the Agro Ecological Zone-8 namely, young Brahmaputra and Jamuna Floodplain. Cultural activities were performed following recommended procedure [**11**]. Fresh jute leaves of different varieties were harvested from 50 days old jute plant for estimation of L-ascorbic acid content. The varieties of *C. olitorius* are O-9897, O-72, O-3820, O-795, MG-1 and Robi-1 and *C. capsularis* are CVL-1, D-154, CC-45, CVE-3, BJC-83 and deshi pat shak-1. Healthy leaves were collected randomly from plants. Three replications were used for each sample. Jute leaves of variety O-9897 and CVL-1 were collected from 56 days old jute plant from the same experimental field for the processing experiment. Ten grams of 56-day-old fresh leaves were processed by -

(i) boiling with de-ionized water for 5, 10 and 15 min,

(ii) soaked in vinegar for 5, 15and 30min,

(iii) spraying vinegar for 5, 15and 30min,

(iv) dried in convection oven in brown paper envelop at 60°C for 2h and

(v) dried in refrigerator in brown paper envelop at 1-8°C for one month.

Leaves of O-9897 and CVL-1 variety of 56 days old jute plant were also prepared as tea and soup and preserved. The drying and preservation methods were as follows -

(i) Tea were prepared by washing, chopping the leaves and then immersion in boiling water for 3min following drying in oven at 60°C afterwards preserved in glass bottle at room temperature,

(ii) Soup were prepared by drying leaves in oven at 100°C overnight, then ground by mortar and pestle, kept in plastic bottle and preserved in freezer (-20°C)

(iii) soup were prepared by drying of leaves in refrigerator at 4°C for 10 days and ground with mortar and pestle, kept in plastic bottle and preserved in freezer at -20°C. L-ascorbic acid content in jute leaves samples was estimated according to the method of Riemschneider et al **(1976)** [**12**].

Ten gm leaves were chopped and homogenized with 50 ml acetic acid (10%) and transferred to a 100 ml volumetric flask and was diluted up to the mark by 10% acetic acid. Then filter through Whatman-no. 1 filter paper. Filtrations were performed in cold at 4°C. Filtrate sample were collected, and two/three drops of Bromine water were added for oxidation of ascorbic acid to dehydroascorbic acid. Then two or three drops of thiourea solution (10%) were added to it to remove excess Bromine water. Then two/three drops of 2, 4 Dinitrophenyle Hydrazine solution were added and kept the sample at 37°C for 3h. Afterwards two or three drops

85% H_2SO_4 was added to form red colour complex. Then absorbance was measured at 280nm using spectrophotometer.

2.2.4.2.2.1 Preparation of standard ascorbic acid solution50mg standard crystalline L ascorbic acid was dissolved in 100 ml of 5% metaphosphoric acid and 10% acetic acid solution. This solution was used for calibration curve preparation. The calibration curve was constructed by plotting the concentration versus the corresponding absorbance.

2.2.4.2.2.2 Statistical analysis Statistical analysis of variance (ANOVA) was carried out using statistical package Minitab to determine variation between treatments. The Least Significant Difference (LSD) test was used for comparison of mean.

2.2.4.2.2.3 Results [Aleya et al 2021, Aleya et al 2022]

L-ascorbic acid content in fresh jute leaves of different varieties of *C. olitorius* and *C. capsularis* are presented in **Table 2.2.4.2.2.1** Different varieties show variation in L-ascorbic acid content in both species. O-9897, O-72 and Robi-1 varieties of *C. olitorius* contain highest L-ascorbic acid (0.61, 0.64 and 0.59 g/10g) than other *C. olitorius* varieties. There is no significant difference among these three varieties. Variety O-3820 and MG-1 contain the second highest L-ascorbic acid (0.49 and 0.51g/10g) and have no significant difference between these two varieties. Variety O-795 contain lowest amount of L-ascorbic acid 0.39g/10g among the six *C. olitorius* varieties. In *C. capsularis* Deshi pat shak-1 contain highest amount of L-ascorbic acid (0.92g/10g). Variety CVL-1 shows the second highest L-ascorbic acid content 0.72g/10g. Variety CVE-3 and CVL-1 did not show significant difference in L-ascorbic acid content. Variety D-154, BJC-83 and CC-45 have less amount of L-ascorbic acid than other *C. capsularis* varieties. Variety CC-45 shows lowest amount of L-ascorbic acid (0.64g/10g) among the six *C. capsularis* varieties. *C. capsularis* contain more L-ascorbic acid than the *C. olitorius*. Variability in L-ascorbic acid content is more in *C. capsularis* species than in *C. olitorius*.

Table 2.2.4.2.2.1 L-ascorbic acid content in fresh leaves of commercially cultivated jute species. [Aleya et al 2021, Aleya et al 2022]

Species	Variety	L-ascorbic acid content(g/10g)[a]
Corchorus olitorius	O-9897	0.61±0.020a
	O-72	0.64±0.021a
	O-3820	0.49±0.021b
	O-795	0.39±0.038c
	MG-1	0.51±0.020b
	Robi-1	0.59±0.010a
Corchorus capsularis	CVL-1	0.72±0.015b
	D-154	0.65±0.015cd
	CC-45	0.64±0.010d
	CVE-3	0.69±0.020bc
	BJC-83	0.65±0.15cd
	Deshi pat shak-1	0.92±0.017a

[a]Mean within a column for L-ascorbic acid content separated with LSD and Values in the column Followed by the same letter are not significantly different, $P < 0.01$

L-ascorbic acid content in different processed jute leaves is presented in **Table 2.2.4.2.2.2** Very few amounts of L-ascorbic acid were lost in leaves of *C. olitorius* during 15 min boiling. Retention of L-ascorbic acid was about 98% during 5-10 min boiling and retention was about 95% by 15min boiling. L-ascorbic acid was increased in leaves of *C. olitorius* while soaked in vinegar for 5min but over time L-ascorbic content was decreased. L-ascorbic was increased 17% by 5min and decreased 26% by 15min and 43% by 30min in vinegar-soaked leaves. Treatment with spraying vinegar on leaves increased L-ascorbic acid content over time during 5-30min treatment. L-ascorbic acid content was increased about 9% by 5min, 75% by 15 min and 97% by 30min of vinegar spraying treatment. Oven dried leaves of *C. olitorius* losses L-ascorbic acid content about 62% at 60°C for 2h. Leaves left in refrigerator at 4°C forone month losses L-ascorbic acid 88%.

Table 2.2.4.2.2.2 Effect on L-ascorbic acid content in leaves of cultivated jute species subjected to 11 processing treatments. [Aleya et al 2021, Aleya et al 2022]

Species (variety)	Processing method	L-ascorbic acid content (g/10 g)
	Fresh leaves (control)	0.81±0.020d
C. olitorius (O-9897)	Boil with H_2O for 5min	0.80±0.026d
	Boil with H_2O for 10 min	0.80±0.017d
	Boil with H_2O for 15min	0.77±0.021d
	Soaked in vinegar for 5 min	0.95±0.029c
	Soaked in vinegar for 15 min	0.55±0.011e
	Soaked in vinegar for 30 min	0.46±0.026f
	Spraying with vinegar for 5 min	0.89±0.015c
	Spraying with vinegar for 15 min	1.42±0.030b
	Spraying with vinegar for 30 min	1.60±0.011a
	Dried at 60°C for 2hrs in oven	0.30±0.010g
	Dried at 4°C for 30 days	0.09±0.010h
	Fresh leaves (control)	1.27±0.020d
C. capsularis (CVL-1)	Boil with H_2O for 5min	0.98±0.010e
	Boil with H_2O for 10 min	0.92±0.010f
	Boil with H_2O for 15min	0.79±0.015g
	Soaked in vinegar for 5 min	1.61±0.020b
	Soaked in vinegar for 15 min	0.79±0.020g
	Soaked in vinegar for 30 min	0.74±0.010h
	Spraying with vinegar for 5 min	1.36±0.010c
	Spraying with vinegar for 15 min	1.59±0.015b
	Spraying with vinegar for 30 min	1.70±0.021a
	Dried at 60°C for 2hrs in oven	0.48±0.026i
	Dried at 4°C for 30 days	0.13±0.021j

[a]Mean within a column for L-ascorbic acid content separated with LSD and Values in the column followed by the same
Letter are not significantly different, $P < 0.01$.

In *C. capsularis* leaves L-ascorbic content was reduced during boiling. The amount of L-ascorbic acid was lost due to boiling about 22%, 27% and 37% by 5, 10 and 15min respectively. That means retention of L-ascorbic acid was 77%, 72% and 62% by 5, 10 and 15min boiling. Vinegar-soaked leaves of *C. capsularis* showed increasing L-ascorbic acid 26% by 5min but decreasing trend was observed over time. Ascorbic content was lost 37% and 41% while leaves soaked in vinegar for15 and 30min. Leaves treated with spraying vinegar increased availability of

L-ascorbic acid 7%, 25% and 33% after 5, 15 and 30min spraying of vinegar, respectively. L-ascorbic acid content was lost 62% in oven dried leaves for 2h at 60°C and about 89% lost in cold dried leaves at 4°C for one month in Refrigerator. During 15min cooking/boiling leaves of *C. olitorius* showed comparatively less reduction than leaves of *C. capsularis*. Leaves of *C. olitorius* variety O-9897 retain L-ascorbic acid 95% (0.77g/10g) of its fresh condition and *C. capsularis* variety CVL-1 retains 62% (0.79g/10g) of its fresh condition. Therefore, after 15 min cooking nearly similar amount of L-ascorbic is available from the leaves of both species. Highest L-ascorbic content was obtained in vinegar spraying treated *C. capsularis* variety CVL-1 (1.7g/10g). The lowest amount of L-ascorbic acid was obtained in one month old cold dried leaves of *C. olitorius* variety O-9897(0.09g/10g). Jute leaves were dried as tea and soup following preservation. Tea prepared fromJute leaves retains very few amounts of L-ascorbic acid. Tea made by leaves of *C. capsularis* contains more amount of L-ascorbic acid than *C. olitorius*. Soup made by cold drying leaves (4°C) retain more L-ascorbic acid than oven drying (60°C) leaves in *C. olitorius* variety O-9897. In leaves of *C. capsularis* variety CVL-1, soup made by oven drying shows more L-ascorbic acid than drying leaves at 4°C in refrigerator.

Table 2.2.4.2.2.3 Effect on L-ascorbic acid content in leaves of cultivated jute species subjected to drying and preservation Methods. [Aleya et al 2021, Aleya et al 2022]

Product	Processing & preservation	Time (Year)	L-ascorbic acid content (g/10g)	
			C. olitorius (O-9897)	*C. capsularis* (CVL-1)
Used as tea	Immersed in boiled water for 3min then cool and then dried at 60°C afterwards preserved in glass bottle in room temperature	1	0.05±0.005	0.07±0.001
Used as soup	Drying in 60°C & preserved in plastic bottle in -20°C	3	0.07±0.001	0.22±0.005
	Drying in 4°C for 10 days & preserved in plastic bottle in -20°C	3	0.14±0.005	0.17±0.008

2.2.4.2.2.4 Discussion

Jute leaves contain good amount of L-ascorbic acid. There is difference among the varieties of two species and between the two species as well. Leaves of *C. capsularis* contain more L-ascorbic acid than leaves of *C. olitorius*. Taste of leaves is different between species. *C. capsularis* taste is bitter and genetic makeup also different [**13**]. All the varieties of both species showed higher amount of L-ascorbic acid than the previous report (**Table 2.2.4.2.2.4**) except one

[**14**]. The present research reports are like the report of Shaker [**14**]. Vitamin C levels may vary considerably due to different factors, which include species, maturity, portion, soil, climate, season, handling, method of preparation and consumption [**15**].

Table 2.2.4.2.2.4 Previous report of L-ascorbic acid content in *Corchorus* sp.[Aleya et al 2021, Aleya et al 2022]

Name	L ascorbic acid content	Reference
Corchorus olitorius L.	316.80 mg/100gm (dw)	[16]
Corchorus olitorius L.	153.63 mg/100ml	[4]
Corchorus olitorius (Serrated edge)	89.94 mg/100gm 101.91 mg/100gm	[17]
Corchorus olitorius (Smooth edge)	40.21 mg/100gm (CV) 44.74 mg/100gm (NBS)	[17]
Corchorus olitorius	7.5mg/gm	[14]
Corchorus capsularis	135.60 mg/100gm	[5]

L-ascorbic acid is water soluble and heat labile. So due to boiling L-ascorbic acid content was reduced. Loss of L-ascorbic acid content differs during boiling of leaves in different species. Leaves of CVL-1 of *C. capsularis* show more loss of L-ascorbic acid than leaves of O-9897 of *C. olitorius*. Leaves of O-9897 retain L-ascorbic acid 95% and CVL-1 retains 62% by 15min boiling. Musa and Obgadoyi (**2012**) revealed that vitamin C concentration decreased significantly in the cooked *C. olitorius* leaves [**18**]. They reported percentage loss of vitamin-C in leaves of *C. olitorius* cooked for 5 and 10min were 79.86% and 85.30%, respectively. The extent of loss of ascorbic acid depends upon variation in cooking methods and periods [**10**]. Shorter cooking time results in greater retention of vitamin during cooking [**19-20**].Fifty percent ascorbic acid retention was achieved in spinach(*Spinacia oleracea*) cooked for 7min[**21**]. Whereas boiling for 6 min caused 64.45% decrease in broccoli (*Brassica oleracea*var. italica), 70.74% in white cabbage (*Brassica oleracea*. var. Caplitata) and 66.82% in cauliflower(*Brassica oleracea*, var. Botrytis) [**22**] which indicates retention of varying ascorbic acid after cooking from species to species. Vinegar may use for pickling vegetables, dressing on salad, preparing for sauce, sanitizing vegetables and help for removal of pesticide residue. Present experimental results showed that vinegar had effect on L-ascorbic acid in leaves. While leaves soaked in vinegar initially L-ascorbic acid was released and then it decreased. When vinegar gets contact with leaf, the leaf tissue may be disintegrated immediately which causes releasing L-ascorbic acid. Afterwards vinegarreacts with L ascorbic acid and break down it. Therefore, in 5min vinegar-soaked leaves released more L-ascorbic acid than its fresh one and then by 15 to 30min L-ascorbic acid were reduced. Similar trend was observed in both species. Spraying treatment of vinegar had different effect, it releases L-ascorbic acid from leaves gradually and continue to release during 5 to 30min treatment. Spraying treatment given with less amount of vinegar compared to leaf which may act slowly on leaves for releasing L-ascorbic acid and over time more L-ascorbic acid was released. Similar trend was observed in both species by spraying vinegar treatment. Depending on the composition of food, the material is prone to nutrient degradationduring drying [**23-24**]. Oven dried leaves of *C. olitorius* and *C. capsularis* lost about 62% L-ascorbic acid and cold drying for one-month leaves losses 88-89% L-ascorbic acid. Making tea and soup from jute leaves were subjected to drying and preservation. Tea made from jute leaves showed lower L-ascorbic acid in both species. Soup prepared from jute leaves showed slightly higher L-ascorbic acid than tea. Tea was preserved in room temperature which may cause deterioration of vitamin content quickly. Soup preserved in freezer. So, it retains slightly more L-ascorbic acid content. Temperature, moisture content and interaction with water all influence the chemistry and biochemistry of the food product during drying and storage [25]. The presence of water in dehydrated foods is important because it affects deterioration reactions and vitamin degradation [**26**].

2.2.4.2.2.5. Conclusion

Jute leaves contain good amount of L-ascorbic acid. Vitamin C or L-ascorbic acid content differ in varieties as well as in between species. *C. capsularis* leaves have more L-ascorbic acid than leaves of *C. olitorius*. Retention of L-ascorbic acid in boiled leaves for 15min was 95% in *C. olitorius* and 62% in *C. capsularis*. Soaking in vinegar for 5min enhance to release more L-ascorbic acid. Spraying treatment of vinegar released maximum L-ascorbic acid from leaves up to 30min. Both cold dried and oven dried leaves showed loss of L-ascorbic acid content. Preserved tea and soup made from jute leaves also lost maximum L-ascorbic acid.

2.2.4.2.2.6 References

[1] Saunders M. Recovery Plan for the Endangered Native Jute Species, Corchorus cunninghamii F. Muell in Queensland (2001–2006) Environmental Protection Agency; 2001.

[2] Azuma K, Nakayama M, Koshioka M, Ippoushi K, Yamaguchi Y, Kohata K, Yamauchi Y. Ito H, Higashio H. Phenolic antioxidants from the leaves of Corchorus olitorius L. Journal of Agricultural and Food Chemistry. 1999; 47(10): 3963-6.

[3] Handoussa H, Hanafi R, Eddiasty I, El-Gendy M, El-Khatib A, Linscheid M, Mahran L, Ayoub N. Anti-inflammatory and cytotoxic activities of dietary phenolics isolated from Corchorus olitorius and Vitis vinifera. Journal of Functional Foods. 2013; 5(3): 1204-16.

[4] Dafam DG, Agunu A, Dénou, A, Kagaru DC, Ohemu TI, Ajima U, Damos JN, Okwori VA. Determination of the ascorbic acid content, mineral and heavy metal levels of some common leafy vegetables of Jos, Plateau State (North Central Nigeria). International Journal of Biosciences. 2020; 16(3): 389-96.

[5] Ali MM, Rahman MT, Ahmed T, Rokeya B, Roy B. Advance research on nutraceutical composition of mature jute leaves. International Journal of Recent Innovations in Medicine and Clinical Research. 2020; 2(4): 124-37.

[6] Szeto YT, Tomlinson B, Benzie IF. Total antioxidant and ascorbic acid content of fresh fruits and vegetables: implications for dietary planning and food preservation. British Journal of Nutrition. 2002; 87(1): 55-9.

[7] Rekha C, Poornima G, Manasa M, Abhipsa V, Devi JP, Kumar HT, Kekuda TR. Ascorbic Acid, total phenol content and antioxidant activity of fresh juices of four ripe and unripe citrus fruits. Chemical Science Transactions. 2012; 1(2): 303-10.

[8] Kaviarasan S, Naik GH, Gangabhagirathi R, Anuradha CV, Priyadarsini KI. In vitro studies on antiradical and antioxidant activities of fenugreek (Trigonella foenum graecum) seeds. Food Chemistry. 2007; 103(1): 31-7.

[9] Li Y, Schellhorn HE. New developments and novel therapeutic perspectives for vitamin C. The Journal of Nutrition. 2007; 137(10): 2171-84.

[10] Bembem K, Sadana B. Effect of cooking methods on the nutritional composition and antioxidant activity of potato tubers, International Journal of Food and Nutritional Sciences. 2013; 2(4): 26–30.

[11] Chowdhury MA, Hassan MS. Handbook of agricultural technology. Bangladesh Agricultural Research Council, Farmgate, Dhaka. 2013 May; 230.

[12] Riemschneider R, Abedin MZ, Mocellin RP. Quality and stabilization testing of heat-preserved foods using Vit C as a criterion medium 1. Alimenta. 1976; 15: 171.

[13] Islam MS, Saito JA, Emdad EM, Ahmed B, Islam MM, Halim A, et al. Comparative genomics of two jute species and insight into fibre biogenesis. Nature. Plants. 2017; 3(2): 16223.

[14] Shaker KN. Effect of micronutrients on growth and antioxidant activity of Corchorus olitorius. Asian Journal of Plant Sciences. 2021; 20(2): 344-54.

[15] Alam MA. Comparative study of total vitamin C in various fruits and vegetables of greater Sylhet area [MSc thesis]. Shahjalal University of Science and Technology, Sylhet. 1996.

[16] Adeniyi SA, Ehiagbonare JE, Nwangwu SCO. Nutritiional evaluation of some staple leafy vegetables in Southern Nigeria. International Journal of Agricultural and Food Science. 2012; 2(2): 37-43.

[17] Ogunlesi M, Okiei W, Azeez L, Obakachi V, Osunsanmi M and Nkenchor G. Vitamin C contents of tropical vegetables and foods determined by voltametric and titrimetric methods and their relevance to the medicinal uses of the plants. International Journal of Electrochemical Science. 2010; 5: 105-15.

[18] Musa A, Ogbadoyi EO. Effect of cooking and sun drying on micronutrients, antinutrients and toxic substances in Corchorus olitorius (Jute Mallow). Journal of Nutrition and Food Science. 2012; 2(3):1-8.

[19] Babalola DA, Makinde YO, Omonona BT, Oyekanmi MO. Determinants of postharvest losses in tomato production. Journal of Life and Physical Science. 2010; 3(2): 14-8.

[20] Zhang D, Hamauzu Y. Phenolics, ascorbic acid, carotenoids and antioxidant activity of broccoli and their changes during conventional and microwave cooking. Food Chemistry, 2004; 88(4): 503-9.

[21] Klein BP, Kuo CHY, Boyd G. Folacin and ascorbic acid retention in fresh raw, microwave, and conventionally cooked spinach. Journal of Food Science. 1981; 46 (2): 640-1.

[22] Shams El-Din MHA, Abdel Kader MM, Makhlouf SK, Mohamed OSS. Effect of some cooking methods on natural antioxidants and their activities in some Brassica vegetables. World Applied Sciences Journal. 2013; 26(6): 697–703.

[23] Achanta S, Okos MR, Cushman JH, Kessler DP. Moisture transport in shrinking gels during saturated drying. AIChE Journal. 1997; 43(8): 2112–22.

[24] Chou SK, Chua KJ. New hybrid drying technologies for heat sensitive foodstuffs. Trends in Food Science &Technology. 2001; 12: 359–69.

[25] Chieh C. Water Chemistry and Biochemistry. In: Hui YH. Eds. Food Biochemistry & Food Processing. USA: Blackwell Publishing. 2006.

[26] Osuna-Garcia JA, Wall MM. Prestorage moisture content affects color loss of ground paprika (Capsicum annuum L.) under storage. Journal of Food Quality. 2007; 21(3): 251-9.

2.2.4.2.3 β-Carotene in jute leaf [Aleya et al 2021, Aleya et al 2022]

2.2.4.2.3.1 Introduction

Jute (*Corchorus* sp.) leaves consumed as vegetable in many countries in Asia and Africa [**1-3**]. Edible species of Corchorus are very good source of proteins, vitamins (A, C, E), mineral nutrients like calcium, iron and carotenoid like lycopene [**4-6**]. Previous research mentioned that jute leaves are rich source of beta carotene [**7, 8**]. Beta carotenes present as micro-nutrients in fruits and vegetables and are responsible for their bright orange, red and yellow colours. In the plant carotenoids act as photosynthetic accessory pigments and also play a protective function as scavengers of oxygen radicals released from chloroplasts during photosynthesis, thus protecting cellular constituents such as DNA from free radical damage [**9, 10**]. One of the most physiological functions of carotenoids in human nutrition is to act as pro-vitamin A (vitamin A precursors). Many studies show strong correlations between carotenoids intake and reduced risk of some diseases, such as cancers [**11**], artherogenesis [**12,13**], bone calcification [**14**], eye degeneration [**15,16**] immune function [**17-19**] and neuronal damage [**20**]. Based on the epidemiological studies a positive link is suggested between higher dietary intake and tissue concentration of carotenoids and lower risk of chronic diseases [**21,22**]. Carotenoids may play a key role as free-radical scavenger and antioxidants in human body tissues [**23, 24**]. B-carotene belongs to the carotene class, which is one of the most abundant carotenoids found in the diet. Fruits and vegetables constitute major source of carotenoids in human diet [**21, 22**]. Nutrition rich vegetables get attention in some region like Sub-Saharan Africa to combat malnutrition and famine. Among other vegetables, *Corchorus* spp. has been identified as promising for up-scaling and addressing food security [**25**]. Therefore, for the perspective of food security, usefulness and nutritional value of indigenous leafy vegetables like jute is required to be rediscovered. There is growing demand of jute leaves for up scaling as its well-known health benefit. Rediscovering nutritional value of jute leaves may be helpful to promote market potentiality as well as awareness among common people. Extraction and estimation procedure of carotenoids are rather complex. High performance liquid chromatography (HPLC) is commonly used for determination of β-carotene. HPLC requires expensive instrument and skillful operation. This work was aimed to investigate the β-carotene content of commercially cultivated jute leaves by simple spectrophotometry and thin layer chromatography (TLC).

2.2.4.2.3.2 Material and methods [Aleya et al 2021, Aleya et al 2022]

2.2.4.2.3.2.1 Plant sample

Jute leaves of nine varieties of Corchorus olitorius and ten varieties of Corchorus capsularis were collected from the experimental field of Breeding Division of Bangladesh Jute Research Institute (BJRI), Dhaka, Bangladesh. The site belongs to the Agro Ecological Zone-8 namely, young Brahmaputra and Jamuna Floodplain. Cultural activities were performed according to Chowdhury and Hassan [**26**]. Fresh healthy jute leaves of selected varieties were harvested from 75 days old jute plant for estimation of β-carotene content. Three biological replications were used for each sample.

2.2.4.2.3.2.2 Estimation of β-carotene using Spectrophotometry

Spectrophotometric method for estimation of β-carotene was followed according to Karnjanawipagul et al [**27**]. β-carotene from jute leaves samples was extracted by the procedures described by Herrero-Martinez et al [**28**] with little modification. Briefly, jute leaves were washed with de-ionized water and chopped. Leaves (10g) were mixed with anhydrous Na_2CO_3 (1g) and homogenized with blender. Homogenized leaves (10g) were transferred into a centrifuge tube and Tetrahydrofuran 20ml was added in it then kept it in the refrigerator overnight. The mixture was centrifuged at 5000rpm for 5min by the centrifuge and supernatant was collected. Dichloromethane 15ml was added to the collected supernatant and then NaCl (10%) ($^{w}/_{v}$) 15ml was added to it and shaken for two min. The extraction was repeated twice, and organic layer was collected in evaporating bowl and placed in the refrigerator (4°C) for 2/3days for evaporation in cold. The residue was collected and kept at -20°C. Afterwards reconstitute with 5ml DCM and diluted ($^{1}/_{40}$-fold) with DCM prior measurement by spectrophotometer. β-carotene samples were measured at 461nm.

2.2.4.2.3.2.4. Detection of β-carotene using thin layer chromatography

β-carotene was detected by thin layer chromatography using the same sample extracted previously for spectrophotometric measurement. The sample was preserved at -20°C and used for TLC work afterwards. An accurately weighted Trans β-carotene standard was dissolved in DCM to obtain a solution of 100μg/ml. Aliquot of standard solution was applied on TLC plate to compare with jute leaves extract of β-carotene. Silica gel on Al foil was stationary phase and hexane was mobile phase. Capillary pipette was used for placing samples on the TLC plate. Silica gel Al foil sheet were cut as per requirements such as 10cm×8.5cm for sample O1-O5, 10cm×7.5cm for sample O6-O9, 10cm×9cm for sample C1-C5 and 10cm×9cm for sample C6 –C10. A horizontal line was drawn by pencil about 1.2cm above from the bottom of the TLC sheet. Capillary tube was dipped in the sample and gently touched it on the pencil drawn line of the TLC plate to empty it. Several pipettes were used for every individual sample. Several short touches were given toempty capillary, so the spot was 1-2mm in diameter. Standard Trans β-carotene (100µg/ml) was used on every plate as standard to compare with jute leaf extract β-carotene. Spots were allowed to dry. Afterwards the TLC sheet with spotted end was placed in a beaker in the developing solvent (hexane). The start spotted line was above the solvent level. Then beaker was covered with glass plate. Solvent was allowed to rise up the TLC plate leaving undisturbed until about 80% of TLC plate is wet. The TLC plate was then removed and taken photo immediately.

2.2.4.2.3.2.5. Preparation of standard β-carotene solution

Standard Trans β-carotene weighted exactly 0.4mg (400µg) was dissolved in 100ml DCM to obtain 4µg/ml β-carotene solution. This solution was used for calibration curve preparation. The calibration curve was constructed by plotting the concentration versus the corresponding absorbance (**Fig2.2.4.2.3.1**).

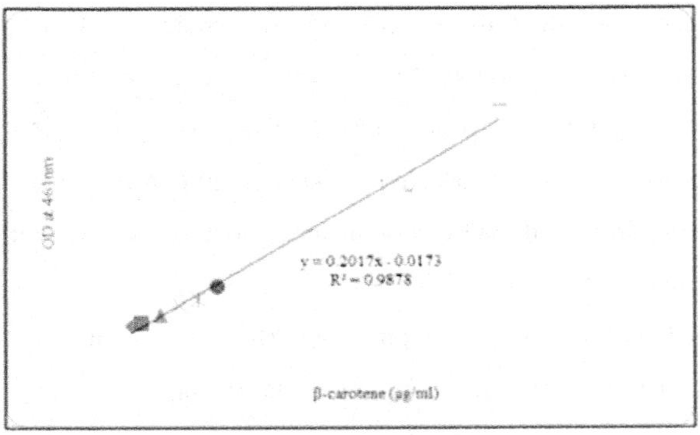

Figure 2.2.4.2.3.1 Calibration curves of β-carotene at 0.1 to 4.0µg/ml. [Aleya et al 2021, Aleya et al 2022]

2.2.4.2.3.3 Results [Aleya et al 2021, Aleya et al 2022]

2.2.4.2.3.3.1 Estimation of β-carotene by spectrophotometry

β-carotene content in jute leaves of different varieties of *C. olitorius* and *C. capsularis* were presented in **Table 2.2.4.2.3.1**. Leaves of varieties of *C. olitorius* contain β-carotene 12.55mg/100g to 39.93 mg/100g. VarietyRobi-1 contain highest amount of β-carotene 39.33mg/100g. Variety O-72, O-4 and O-3820 contain the next highest 30.46, 29.05 and 28.93 mg/100g β-carotene respectively after Robi-1. Variety O-9897, OM-1, JRO-524 and MG-1 have medium amount, 25.67, 24.41, 22.93 and 20.22mg/100g β-carotene and O-795 contain lowest amount of β-carotene 12.55mg/100g in comparison to other *C. olitorius* varieties. β-carotene ranged 11.34 to 22.28mg/100g in leaves of different varieties of *C. capsularis*. Among the *C. capsularis* varieties CVL-1 showed the highest amount of β-carotene 22.28mg/100g. Variety D-154 (21.79mg/100g), BJC-2197 (21.21mg/100g) and Deshi pat shak-1 (20.66mg/100g) is comparable to variety CVL-1 having almost same level of β-carotene. Variety BJC-5003 and BJC-7370 contain 19.18mg/100g and 18.08mg/100g β-carotene. BJC-83 and BJC-2142 have lowest β-carotene 11.34mg/100g and 12.32mg/100g respectively.

Table 2.2.4.2.3.1. Beta carotene content of leaves ofdifferent *Corchorus olitorius* and *Corchorus capsularis* varieties. [Aleya et al 2021, Aleya et al 2022]

Species	Variety	Beta carotene conc. mg/100gm leaves
Corchorus olitorius	O-3820	28.93±0.54
	O-795	12.55±0.46
	O-4	29.05±0.42
	O-72	30.46±0.61
	O-9897	25.67±0.57
	OM-1	24.41±0.38
	MG-1	20.22±0.61
	JRO-524	22.93±0.60
	Robi-1	39.93±0.49
Corchorus capsularis	CVL-1	22.28±0.59
	CVE-3	15.87±0.42
	D-154	21.79±0.42
	CC-45	17.54±0.60
	BJC-2142	11.34±0.34
	BJC-83	12.32±0.44
	BJC-7370	18.08±0.51
	BJC-5003	19.18±0.49
	BJC-2197	21.21±0.35
	Deshi pat shak-1	20.66±0.50

2.2.4.2.3.3.3.2 Detection of β-carotene in jute leaves by TLC [Aleya et al 2021, Aleya et al 2022]

Figure 2.2.4.2.3.2 represents the thin layer chromatographic results of β-carotene in different varieties of *C. olitorius* leaves. **Figure 2.2.4.2.3.3** represent the β-carotene in different varieties of *C. capsularis* leaves.

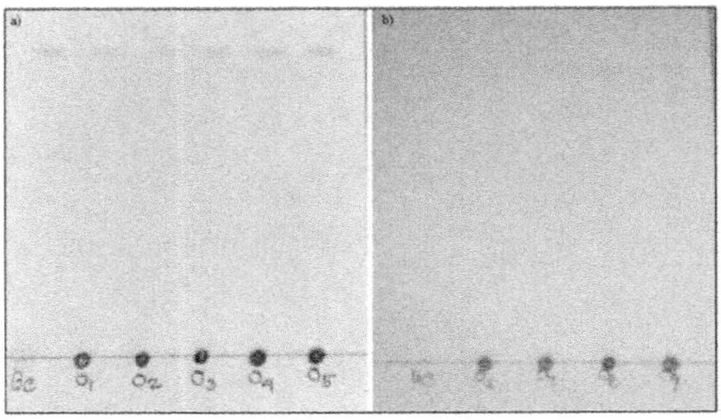

Figure 2.2.4.2.3.2 β-carotene detection of nine varieties of *C. olitorius* leaves extract by TLC a) Bc, standard β-carotene; O 1 , O-795; O 2 , O-9897; O 3 , O-72; O 4 , O-4; O 5 , OM-1 b) Bc, standard β-carotene; O 6 , O-3820; O 7 , MG-1; O 8 , Robi-1; O 9 , JRO-524. [Aleya et al 2021, Aleya et al 2022]

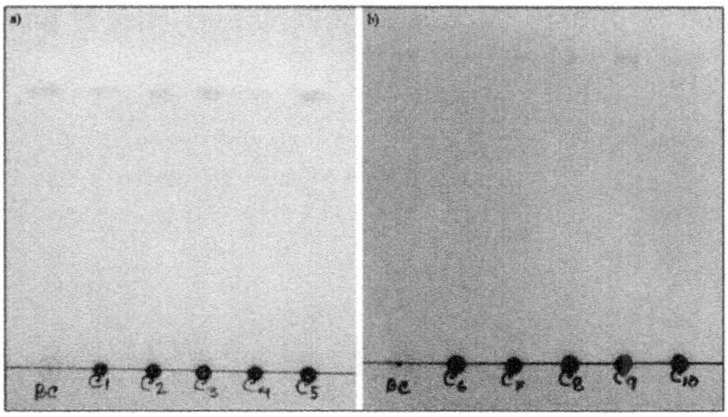

Figure 2.2.4.2.3.3 β-carotene detection of ten varieties of *C. capsularis* leaves extract by TLC a) Bc, standard β-carotene; C 1 , CVL-1; C 2 , D-154; C 3 , BJC-7370 C 4 , CC-45; C 5 , BJC-2197 b) Bc, standard β-carotene; C 6 , Deshi pat shak-1; C 7 , BJC-83; C 8 , BJC-2142; C 9 , BJC-5003; C 10 , CVE-3. [Aleya et al 2021, Aleya et al 2022]

It is observed in every TLC plate that β-carotene of the leaf sample reached to the similar position as the standard β-carotene did and orange colour of all the samples were distinct. Rf value of the jute leaves β-carotene is 0.82 in **Table 2.2.4.2.3.2** (TLC plate-1) and 0.85 in **Table 2.2.4.2.3.2**(TLC plate-2, 3and 4). R_x value of jute leaves β-carotene was 1.

Table 2.2.4.2.3.2 R_f and R_x of Beta carotene of different varieties of jute leaves in TLC plate. [Aleya et al 2021, Aleya et al 2022]

Jute species	TLC plate (Figure)	Sample designation	Rf	Rx
C. olitorius	Plate 1 (Fig. 2a)	Bc (standard β-carotene)	0.82	
		O_1 (O-795)	0.82	1
		O_2 (O-9897)	0.82	1
		O_3 (O-72)	0.82	1
		O_4 (O-4)	0.82	1
		O_5 (OM-1)	0.82	1
	Plate 2 (Fig. 2b)	Bc (standard β-carotene)	0.85	
		O_6 (O-3820)	0.85	1
		O_7 (MG-1)	0.85	1
		O_8 (Robi-1)	0.85	1
		O_9 (JRO-524)	0.85	1
C. capsularis	Plate 3 (Fig. 3a)	Bc (standard β-carotene)	0.85	
		C_1 (CVL-1)	0.85	1
		C_2 (D-154)	0.85	1
		C_3 (BJC-7370)	0.85	1
		C_4 (CC-45)	0.85	1
		C_5 (BJC-2197)	0.85	1
	Plate -4 (Fig. 3b)	Bc (standard β-carotene)	0.85	
		C_6 (Deshi pat shak-1)	0.85	1
		C_7 (BJC-83)	0.85	1
		C_8 (BJC-2142)	0.85	1
		C_9 (BJC-5003)	0.85	1
		C_{10} (CVE-3)	0.85	1

2.2.4.2.3.4 Discussion

β-carotene content shows variation in different varieties of jute leaves. Most of olitorius leaves contain higher β-carotene than the β-carotene content in leaves of *C. capsularis*. In some varieties of *C. olitorius* shows similarity with some*C. capsularis* varieties. MG-1 (20.22mg/100g) and JRO-524 (22.93) of *C. olitorius* varieties shows β-carotene content like varieties CVL-1(22.28mg/100g), D-154 (21.79mg/100g), BJC-2197 (21.2mg/100g) and deshi pat shak-1 (20.66mg/100g) of capsularis varieties. Olitorius variety O-795 shows β-carotene content (12.55mg/100g) like capsularis varieties BJC-2142 (11.34mg/100g) and BJC-83 (12.32mg/100g). Choudhary and his colleagues [**29**]mentioned that *C. olitorius* showed the highest amount of β-carotene among the 17 genotypes belonging to six jute species. They investigated β-carotene content in different Corchorus species including some varieties of *C. olitorius*. They observed most of the *C. olitorius* varieties have higher amount of β-carotene, some were similar, and some were lower than a capsularis variety (**Table2.2.4.2.3.3**). In this study, jute leaves of different varieties showed higher amount of β-carotene (**Table 2.2.4.2.3.1**) than the previous report presented in **Table2.2.4.2.3.3** [**29-31**]. This may be occurred due to the geographical origin, genetic make-up, climate, soil condition and structure and fertilizers [**32**].

Table 2.2.4.2.3.3 Beta carotene content in jute leaves in previous report. [Aleya et al 2021, Aleya et al 2022]

Species	Condition	β-carotene content	Reference
C. olitorius	Fresh leaves	9774.90μg/100g	Traoré *et al.*, 2019 [30]
	Cooked using ash leachet for 15 and 30min	3604.00μg/100g and 3768.80μg/100g	
	Cooked in water for 15 and 30min	2285.50μg/100g and 2365.00μg/100g	
	Fried for 5min following 30min cooking	86.42%	
C. fascicularis	Fresh leaves	62.67 mg/kg	Choudhary *et al.*, 2013 [29]
C. triloculairs		36.00	
C. aestuans		76.33	
C. tridens		47.00	
C. capsularis (Maniksari)		61.33	
C. olitorius (Sudan Green)		81.33	
C. olitorius (Tanganyika)		79.70	
C. olitoruius (TJ-40)		64.00	
C. olitorius (Bidhan Rupali)		34.33	
C. olitorius (KOM -62)		63.67	
C. olitorius (JRO-66)		67.33	
C. olitorius (JRO-3690)		66.00	
C. olitorius (JRO-8432)		72.67	
C. olitorius (JRO-204)		51.00	
C. olitorius (JRO-524)		74.00	
C. olitorius (JRO-632)		53.33	
C. olitorius (S-19)		50.67	
C. capsularis (CVL-1)	Fresh leaves	7.82±0.16 mg/100g	Ali et al., 2020 [31]
	Shade dried	4.65±0.32	
	Solar dried	4.42±0.28	
	Cooked	3.81±0.42	

Thin layer chromatography results show the presence of β-carotene content in jute leaves. B-carotene from the leaf samples reached to the position which similar to the position of standard trans-beta carotene sample. Orange colour of the spot was visualized immediately in each variety of leaf samples in every TLC plate (**Fig 2.2.4.2.3.2and 2.2.4.2.3.2.3**). β-carotene is bright yellowish-orange due to a system of conjugated double bonds in its structure, it could be identified visually on the chromatogram without using any detecting reagents. The same occurrence observed during this experiment. The spots colours were visualized without any detecting reagent. $R_f$value of experimented samples were calculated high (0.82/85). It was mentioned in previous report that non oxygenated carotenoids like β-carotene have higher R_f value [**33, 34**]. Rx value is calculated 1.0 for all the β-carotene in leaves of different varieties of *C. olitorius* and *C. capsularis* (**Table 2.2.4.2.3.2**). Trans β-carotene was used as standard, so the Rx results show clear indication of presence of Trans β-carotene in fresh jute leaves.Carotenoids are insoluble in water but soluble in organic solvents. They present specific absorption spectra that help for their identification. In contact with air, they self-oxidate quickly and degrade [**35**]. Carotenoids require specific extraction and separation procedures due to its wide range of polarity. β-carotene is practically insoluble in water and hardly soluble even in methanol, acetonitrile or dimethyl sulfoxide [**36-38**]. Solubility is better in hexane, chloroform and dichloromethane (DCM), highest solubility was observed in Tetrahydrofuran (THF). In this experiment THF was used following DCM twice. So, the extraction process may be performed perfectly that may contribute for obtaining better results of β-carotene by spectrophotometer as well as by TLC. Extraction of carotenoids must be performed very quickly, avoiding exposure to light, oxygen and high temperature [**39, 40**]. Nitrogen atmosphere is suggested for extraction of β-carotene to maintain cold environment. Here in this experiment cold environment was maintained in refrigerator. Extracted sample was dried at 4°C in refrigerator for 2/3 days. The procedure followed neither nitrogen atmosphere nor quick procedure as mentioned in protocol, but it worked well. Thin layer chromatography results showed that good extraction was performed. TLC method is applied recently for separation and identification of complicated compounds such as carotenoids [**41, 42**]. In Thin layer chromatography stationary phases are polar and mobile phases are nonpolar or low polarity. Early TLC methods used silica as stationary phase and a nonpolar mobile phase for separation of carotenoids [**43-46**]. It was also suggested that best separations could be achieved by use of adsorption TLC on alumina with hexane-chloroform mixtures as mobile phases [**47**]. In most cases petroleum ether with acetone was used major organic mobile phase [**48-49**]. Hexane was found to be the second major mobile phase used in TLC of carotenoids from plant sources [**50, 51**]. In this TLC experiment silica gel on TLC Al foils were used as stationary phase and hexane was the non-polar mobile phase. B-carotene in jute leaves were detected well by this TLC system. Thin layer chromatography showed potentialityforseparation and identification of carotenoids in vegetable and biological samples [**52**]. Our TLC reports for β-carotene detection of jute leaves are in agreement with this report.

2.2.4.2.3.5 Conclusion

Jute leaves are very rich in β-carotene content. Variation was found in β-carotene in jute leaves of different varieties. Most of the varieties in Corchorus olitorius species showed higher β-carotene than the varieties of Corchorus capsularis. In TLC system silica gel on TLC Al foils as stationary phase and hexane as mobile phase performed good separation and identification of β-carotene in jute leaves.

2.2.4.2.3.6 References

[1] Westphal-Stevels J. Local vegetables in Cameroon: Corchorus species used as a vegetable. Acta Horticulturae. 1986; 182: 423–426.

[2] Kinabo J, Mnkeni AP, Nyaruhucha CNM, Msuya J, Haug A, Ishengoma J. Feeding frequency and nutrient content of foods commonly consumed in the Iringa and Morogoro regions in Tanzania. International Journal of Food Sciences and Nutrition. 2006; 57(1-2): 9-17.

[3] Nyadanu D, Lowor ST. Promoting competitiveness of neglected and underutilized crop species: comparative analysis of nutritional composition of indigenous and exotic leafy and fruit vegetables in Ghana. Genetic Resources and Crop Evolution. 2014; 62(1): 131-140.

[4] Steyn NP, Oliver J, Winter P, Burger S, Nesamvuni S. A survey of wild, green, leafy vegetables and their potential in combating micronutrient deficiencies in rural populations. South African Journal of Science. 2001; 97: 276-279.

[5] Dansi A, Adajatin A, Adoukonou-Sagbadjia H, Faladé V, Yedomonhan H, Odou D, Dossou B. Traditional leafy vegetable and their use in the Benin Repubic. Genetic Resources and Crop Evolution. 2008; 55: 1239-1256.

[6] Nasreen A, Ahmed Z, Ali M, Shimu T. Jute leaves: A potential sources of lycopene. International Journal of Vegetable Sciences. 2022; 28(2): 180-188.

[7] Agbemafle R, Obodai EA, Adukpo GE, Amprako DN. Effects of boiling time on the concentrations of vitamin c and beta-carotene in five selected green vegetables consumed in Ghana. Advances in Applied Science Research. 2012; 3(5): 2815-2820.

[8] Traoré K, Parkouda C, Savadogo A, Hama F, Kamga R, Traoré Y. Effect of processing methods on the nutritional content of three traditional vegetables leaves: amaranth, black nightshade and jute mallow. Food Science and Nutrition. 2017; 5(6): 1139-1144.

[9] Miki W. Biological functions and activities of animal carotenoids. Pure and Applied Chemistry. 1991; 63(10):141-146.

[10] Nishino H, Murakoshi M, Li T, Takemura M, Kuchide M, Knazawa M, Mou XY, Wada S, Masuda M, Ohsaka Y, Yogosawa S, Satomi Y, Jinno K. Carotenoids in cancer chemoprevention. Cancer Metastasis Reviews. 2002; 21: 257-264.

[11] Nishino H. Prevention of hepatocellular carcinoma in chronic viral hepatitis patients with cirrhosis by carotenoid mixture. Recent Results Cancer Research 2007; 174: 64-71.

[12] Lidebjer C, Leanderson P, Ernerudh J, Jonasson L. Low Plasma levels of oxygenated carotenoids in patients with coronary artery disease. Nutrition, metabolism, and cardiovascular diseases. 2007; 17: 448-456.

[13] Osganian SK, Stampfer MJ, Rimm E, Spiegelman D, Manson JE, Willett WC. Dietary carotenoids and risk of coronary artery disease in women. American Journal of Clinical Nutrition. 2003; 77: 1390-1399.

[14] Yamaguchi M, Uchiyama S. Effect of carotenoid on calcium content and alkaline phosphatase activity in rat femoral tissues in vitro: the unique anabolic effect of beta-cryptoxanthin. Biological and Pharmaceutical Bulletin.
2003; 26:1188-1191.

[15] Semba RD, Daganelie G. Are lutein and zeaxanthin conditionally essentials nutrients for eye health medical? Hypothesis. 2003; 61: 465-472.

[16] Schalch W, Chon W, Barker FM. Köpcke W, Mellerio J, Bird AC, Robsn AG, Fitzke FF. van Kuijk FJGM. Xanthophyll accumulation in the human retina during supplementation with lutein or zeaxanthin-the LUXEA (Lutein xanthophylls Eye Accumulation) study. Archives of Biochemistry and Biophysics. 2007; 458:128-135.

[17] Sato Y, Akiyama H, Suganuma H, Watanabe T., Nagaoka MH, Inakuma T, Goda Y, Maitani T. The Feeding of beta-carotene down-regulates serum IgE levels and inhibits the type I allergic response in mice. Biological and Pharmaceutical Bulletin. 2004; 27: 978-984.

[18] Hughes DA. Dietary carotenoids and human Immune function. Nutrition. 2001; 17: 978-984.

[19] Okai Y, Higashi-Okai K. Possible immune modulating activities of carotenoids in in vitro cell culture experiments. International Journal of Immunopharmacology. 1996; 18: 753-758.

[20] Stuerenburg HJ, Ganzer S, Muller-Thomsen T. Plasma beta carotene in Alzheimer's disease. Association with cerebrospinal fluid beta-amyloid 1-40 (Abeta40), beta-amyloid 1-42(Abta42) and total Tau. Neuroendocrinology Letters. 2005; 26: 696-698.

[21] Johnson EJ. The role of carotenoids in human health. Nutrition in Clinical Care. 2002; 5:56-65.

[22] Agarwal S, Rao AV. Carotenoids and chronic diseases. Drug Metabolism and Drug Interactions. 2000; 17:189-210.

[23] Elliott R. Mechanisms of genomic and nongenomic actions of carotenoids. Biochimica et Biophysica Acta. 2005; 1740(2): 147-154.

[24] Bendich A, Olson JA. Biological actions of carotenoids. FASEB Journal. 1989; 3:1927-1932.

[25] Muthoni J, Nyamongo DO, Silingi M. Participatory characterization and evaluation of some African leafy vegetables in Lari, Kiambu West District, Central Kenya. Journal of Horticulture and Forestry. 2010; 2(2): 12–16.

[26] Chowdhury MAH, Hassan MS. Handbook of agricultural technology. Bangladesh Agricultural Research Council, Dhaka, Bangladesh. 2013.

[27] Karnjanawipagul P, Nittayanuntawech W, Rojsanga P, Suntornsuk L. Analysis of β-Carotene in Carrot by Spectrophotometry. Mahidol university Journal of Pharmaceutical Science. 2010; 37(1-2): 8-16.

[28] Herrero-Martinez JM, Eeltink, Schoenmakers PJ, Kok WT, Ramis-Ramos G. Determination of major carotenoids in vegetables by capillary electrochromatography. Journal of Separation Science. 2006; 29: 660-665.

[29] Choudhary SB, Sharma HK, Karmakar PG, Kumar A, Saha AR, Mahapatra BS. Nutritional profile of cultivated and wild jute (Corchorus) species. Australian Journal of Crop Science. 2013; 7(13): 1973-1982.

[30] Traoré K, Parkouda C, Savadogo A, Ba/Hama F, Kamga R, Traoré Y. Effect of processing methods on the nutritional content of three traditional vegetables leaves: Amaranth, black nightshade and jute mallow. Food Science and
Nutrition. 2017; 5(6): 1139-1144.

[31] Ali MM, Rahman MT, Ahmed T, Rokeya B, Roy B. Advance Research on Nutraceutical Composition of Mature Jute Leaves. International Journal of Recent Innovations in Medicine and Clinical Research. 2020; 2(4): 124-137.

[32] Rodriguez-Amaya DB. Carotenoids and food preparation: The retention of provitamin A carotenoids in prepared, processed, and stored foods, Department of Food Science, University of Campinas, C.P. 6121, 13083-970
Campinas, SP., Brazil. 1997.

[33] Mariana G, Corina M, Alina C, Sanda B. Determination of carotenoids by thin layer chromatography. Annals of the University of Oradea, Environmental Protection Issue. 2016; XXVI.

[34] Perucka I, Oleszek W. Extraction and determination of capsaicinoids in fruit of hot pepper Capsicum annuum L. by spectrophotometry and high-performance liquid chromatography. Food Chemistry. 2000; 71(2): 287-291.

[35] Britton G, Liaaen-Jensen S, Pfander H. Carotenoids. vol 4, Natural functions. Birkhäuser Verlag, Basel. 2008; 431.

[36] Feltl L, Pacakova V, Stulik K, Volka K. Reliability of carotenoid analyses: A review. Current Analytical Chemistry. 2005; 1: 93-100.

[37] Aust O, Sies H, Stahl W, Polidori MC. Analysis of lipophilic antioxidants in human serum and tissues: tocopherols and carotenoids. Journal of Chromatography. 2001; 936: 83-93.

[38] Craft NE, Soares JH. Relative solubility, stability, and absorptivity of lutein and β - carotene in organic solvents. Journal of Agricultural Food Chemistry. 1992; 40: 431-434.

[39] Van den Berg H, Faulks R, Granado HF, Hirschberg J, Olmedilla B, Sandmann G, Southon S, Stahl W. The potential for the improvement of carotenoid levels in foods and the likely systemic effects. Journal of the Science of Food and Agriculture. 2000; 80: 880-912.

[40] Marsili R, Callahan D. Comparison of a liquid solvent extraction technique and supercritical fluid extraction for the determination of alpha- and beta-carotene in vegetables. Journal of Chromatography Science. 1993; 31(10): 422-428.

[41] Checheta OV, Safonova EF, Slivkin AI. Thin-layer chromatography characteristics of [beta]-carotene. Pharmaceutical Chemistry Journal. 2012; 46(5): 321.

[42] Kirchner JG, Thin Layer Chromatography, 2nd Ed., Techniques of Chemistry, Vol.14 Somerset, New Jersey, U.S.A. 1978.

[43] Cserhati T, Forgacs E. Liquid chromatographic separation of terpenoid pigments in foods and food products. Journal of Chromatography A. 2001; 936(1-2): 119-137.

[44] Vinkler M, Kiszel-Richter M. A thin layer chromatographic method to determine the pigment content (components) in the pericarb of paprika. Acta Aliment. 1972; 1: 41-58.

[45] Buckle KA, Rahman FMM. Separation of chlorophyll and carotenoid pigments of capsicum cultivars. Journal of Chromatography A. 1979; 171: 385-391.

[46] Cserhati T, Forgacs E, Hollo J. Separation of color pigments of capsicum-annuum by adsorption and reversed-phase thin-layer chromatography. Journal of Planar Chromatography. 1993; 6(6): 472-475.

[47] Forgacs E, Cserhati T. Thin-layer chromatography of natural pigments: new advances. Journal of Liquid Chromatography & Related Technologies. 2002; 25(10-11): 1521-1541.

[48] Ren D, Zhang S. Separation and Identification of the Yellow Carotenoids in Potamogeton crispus L. Food Chemistry. 2008; 106(1): 410-414.

[49] Melendez-Martinez AJ, Britton G, Vicario IM, Heredia FJ. HPLC analysis of geometrical isomers of lutein epoxide isolated from dandelion (Taraxacum officinale F. Weber ex Wiggers). Phytochemistry. 2006; 67(8): 771-777.

[50] Liu H, Kao T, Chen B. Determination of Carotenoids in the Chinese Medical Herb Jiao-Gu-Lan (Gynostemma Pentaphyllum MAKINO) by Liquid Chromatography. Chromatographia. 2004; 60: 411-417.

[51] Rao AR, Sarada R, Ravishankar GA. Stabilization of astaxanthin in edible oils and its use as an antioxidant. Journal of the Science of Food and Agriculture. 2007; 87(6): 957-965.

[52] Zeb A, Murkovic M. Thin-Layer Chromatographic Analysis of Carotenoids in Plant and Animal Samples. Journal of Planar Chromatography. 2010; 23: 94-103.

2.2.4.2.4 Antibacterial properties of jute leaf [Taslima et al 2022]

2.2.4.2.4.1 Introduction

Jute is found in tropical and subtropical regions of Bangladesh and is traditionally used for fiber but it has medicinal applications too. Jute leaf tea has been studied for antibacterial activity. Many research reveals about antibacterial property of jute leaves. Jute leaves (pat shak) are favorite vegetables in Bangladesh. Rural people in Bangladesh think pat shak cures fever. Antibacterial property of jute leaf may support belief of common rural people about anti-fever quality of jute leaf. In our country jute leaves are well known as folk medicine for remedy of fever, removal of hook warm, etc. [**1-3**]. Besides, Bangladesh some other country cultivates jute as herbal medicinal plant.Microbiology and Biochemistry Department of Bangladesh Jute Research Institute (BJRI) has been working on jute leaf tea. As tea jute leaves antibacterial property may study for commercialization of the product. Along with reveal the proof of some herbal medication as jute leaf. Based on the fact that there is no scientific research reporting on the antimicrobial activity of this jute leaf tea, Therefore this study has been undertaken to screen for its potential antibacterial activity. Jute leaf is the good source of vitamin, nutrient, minerals etc. Tea provides antioxidant and some other health benefit also. Vegetables constitute essential components of the meal by contributing protein, vitamins, iron, calcium, and other nutrients. That is usually in shortage of supply in the daily diets. It is important to provide vitamin C, vitamin A and other vitamins also in the diet [**1, 2**]. Rural people preserve jute leaves by drying and use it later for different medicinal purposes. During preservation, some portion of nutrients may loss. To make value added product from jute leaves, level of antimicrobial and antioxidant are required to know. Therefore antimicrobial properties of jute leaves and antioxidant level of processed jute leaves should be analyzed. Research revealed that tea and herbal tea both have antibacterial potentiality. Antibacterial activity of tea may provide some health benefit by controlling bacterial growth and activity.Because disease causes by that specific bacteria may be prevented or cured. Present experiment has been undertaken to investigate the antibacterial property of jute leaf tea. The present study was taken to investigate the antibacterial properties of jute leaf tea, improve the quality of jute leaf tea and to give theinformation about health potentiality of jute leaf tea. This study will help to obtain information about antibacterial activity status of jute leaf tea.

2.2.4.2.4.2. Material and methods [Taslima et al 2022]

Present study was conducted in the department of Microbiology and Biochemistry, Head Office, Bangladesh Jute Research Institute (BJRI), Dhaka using the equipments, chemicals and raw materials where all the chemicals and media were reagent grade (Sigma). The standard bacterial samples (**Table 1**) collected. Fresh jute leaves of some Deshi (*C. capsularis* CVL-1) and Tossa (*C. olitorius* O-9897) were collected form premises of BJRI, Dhaka. Jute leaf tea was made of both deshi (CVL-1) and tossa (O-9897) jute leaves.

2.2.4.2.4.2.1. Extraction of samples from jute leaves The leaf extracts were prepared by sequential coldmaceration method using distilled water as a solvent [**4**]. Fifty grams of dried powder of leaf material was soaked in 250ml distilled water for 24h at room temperature under shaking condition at 120rpm. This solution was filtered with the help of Whatman No. 1 filter paper and the filtrate was collected in sterile flask at room temperature which was further used for antimicrobial assay after dilution.

2.2.4.2.4.2.2. Sample Preparation Fresh leaves of jute were washed with distilled water followed by air drying and then homogenized to fine powder and stored in airtight bottles at 4°C in aseptic condition [**4**].

2.2.4.2.4.2.3. Microorganisms used in the present investigation The antibacterial activity of jute leaf samples were checked on nutrient agar media against different pathogenic and non-pathogenic microorganisms. Different pure culture (ATCC) microbes were procuredfrom the Primeasia University, Dhaka, Bangladesh. Out of all bacteria used in present investigation, *Salmonella typhi, Klebsiella* sp., *Escherichia coli,* and *Pseudomonas* sp. were Gram negative bacteria and Acenetobacter sp. and Bacillus cereus was Gram positive bacterium. All the bacterial strains used in present investigation were maintained at 4°C on nutrient agar medium.

2.2.4.2.4.2.4. Preparation of bacterial suspension Colonies of different strains of bacteria were aseptically transferred to the individual flasks containing fresh nutrient broth and further incubated at 37° C for 24h.

2.2.4.2.4.2.5. Treatments

A. Comparative treatment- Jute leaf of 0.03g and 0.05g jute leaf antibacterial activity against different bacterial isolates.

B. Time duration treatment- Study of Jute leaf (O-9897 and CVL-1) antibacterial activity against different bacterial isolates boiling at different time interval (5, 10, 15 and 20min).

2.2.4.2.4.3 Results and discussion

The tea liquors were prepared using all these tree different treatments and it these liquors antimicrobial activities were tested (**Fig. 2.2.4.2.4.1**). Antibacterial activity of aqueous extracts of jute leaf was evaluated by agar well diffusion method against different pathogenic bacteria (**Fig 2.2.4.2.4.2 and 2.2.4.2.4.3**). In this study, highest antibacterial activity was observed against *E. coli*(ZOI=17.0±1.0 mm) by extract followed by inhibition of Salmonella typhi by aqueous extract (ZOI=14.5±0.5 mm). Overall, it is clear that extract having maximum antibacterial activity against all the test organisms, followed by aqueous extracts. Infectious diseases are the major cause of morbidity and mortality worldwide. The number of multi drug resistant microbial strains and the appearance of strains with reduced susceptibility to antibiotics are continuously increasing due to indiscriminate use of antibiotics and other antibacterial agents in treatment. The plant based traditional medicines were proven highly effective for their utilization as a source of antimicrobial compounds [**5**]. The beneficial medicinal effects of plant materials typically result fromthe secondary metabolites present in the plant, although; it is usually not attributed to a single compound but a combination of the metabolites. Plants are potentially useful for the development of chemotherapeutic agents. The plants have traditionally provided a source of hope for novel drug compounds, as plant herbal mixtures have made large contributions to human health and well being. The use of plant extracts with known antimicrobial properties can be of great significance for therapeutic treatment [**6**]. One of the study showed that the aqueous crude extract of *Emblica officinalis* possess strong inhibitory action against all the test human bacterial pathogens such as *Bacillus* sp., *Lactobacillus* sp., *Pseudomonas* sp., *Proteus* sp. and *Streptococcus* sp. by using agar well diffusion method [**7**]. Dharajiya et al [**8**] screened the in vitro antibacterial activity of aqueous, ethanol and acetone extracts of fruitsof *Emblica officinalis* against Gram-positive versus Gram-negative bacteria employing Staphylococcus aureus and *E. coli*, respectively.

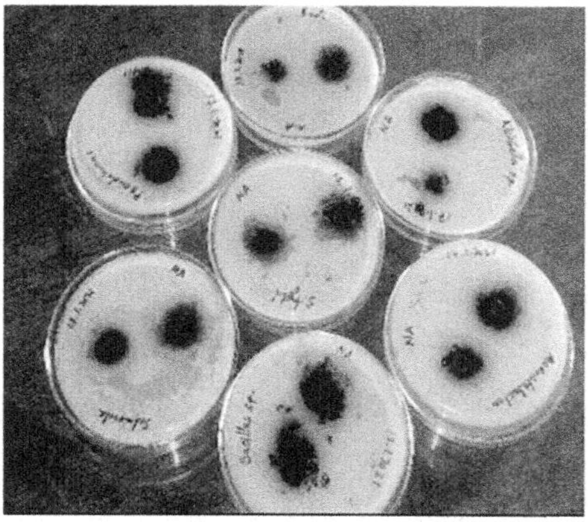

Figure 2.2.4.2.4.1 Jute leaf antibacterial activity against different bacterial isolates (0.03 g and 0.05 g jute leaf) [Taslima et al 2022]

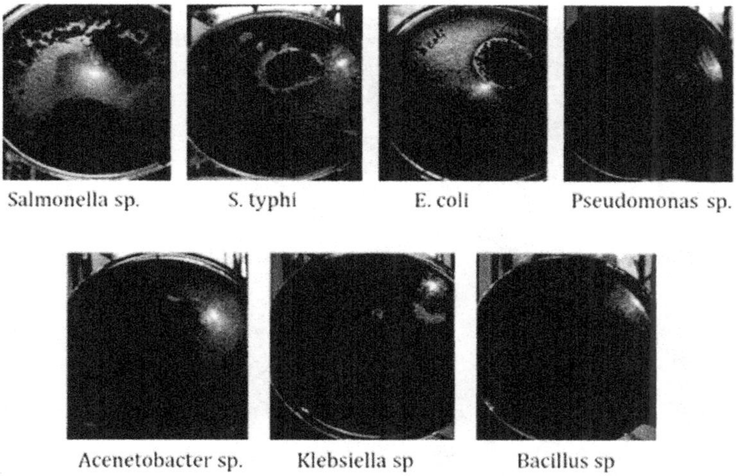

Figure 2.2.4.2.4.2 Jute leaf (0.03 g and 0.05 g jute leaf) antibacterial activity against individual bacterial isolates. [Taslima et al 2022]

Jute leaf has antimicrobial activity against different isolated bacterial sample. All the extracts exhibited significant antibacterial activity, more against *S. aureus* than *E. coli*. *Emblica officinalis* possess potent antibacterial activity against *E. coli, K. ozaenae, K. pneumoniae, Proteus mirabilis, Pseudomonas aeruginosa, S. paratyphi* A, *S. paratyphi* B and *S. marcescens* [**9**]. Plant phytochemical constituents such as saponins, alkaloids, flavonoids, tannin and some other aromatic compounds are secondary metabolites those serve as defence mechanisms against many microorganisms, insects and other herbivores [**10**]. In the qualitative analysis of phytochemicals, the methanol and aqueous extract of E. officinalis fruits showed the presence of alkaloids, tannins, saponins, cardiac glycosides, steroids, phenols, glycoside and flavonoids. While, methanol extract of V. negundo leaves indicated presence of all the phytochemicals analysed in present study, except sterol. Hexane extracts of both the plant showed presence of less phytochemicals as compared toother extracts which might be reason behind their reduced antimicrobial activity. It has been suggested that the antimicrobial activity of the plant is mainly due to the presence of essential oils, flavonoids, terpenoids, alkaloids, tanins, saponins and other natural polyphenolic compounds or free hydroxyl groups in plant extracts [**11**]. Presence of more bioactive phytochemicals in the extract may be the reason behind the higher inhibitory action of the particular extract [**12**].

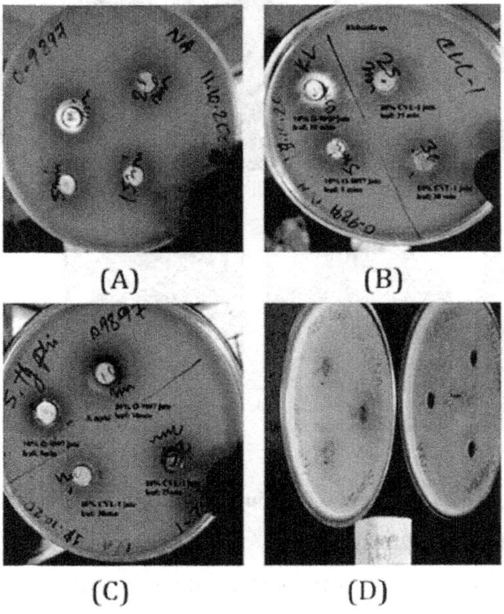

Figure 2.2.4.2.4.3 Comparative study of antibacterial activity of jute leaf extract of O-9897 & CVL-1 at different time interval (5, 10, 15 and 20min) with boiling against (A) *B. cereus*,(B) *Klebsiella* sp. (C) *S. typhi* and (D) *S. aureus*. [Taslima et al 2022]

It has been reported that many volatile components are effective against Gram positive and Gram negative bacteria. The leaves of plant are used commonly by the people in food as a vegetable. The human body is able to directly absorb the different compounds and due to the low concentration of these compounds, it has transpired that the plant particularly provides effective protection against infection. This protective property, either through the plant itself or by the compounds contained within, also has the effect of prolonging the shelf life of food in which it is used or prevents it being spoilt by microorganisms. Its current use as a plant additive to other food and consumed by the public may be due to its health-giving properties and its taste. There is a definite need for more research to determine the metabolite synthesis, other antimicrobial properties that occur in this jute leaf, as well as the construction of new molecules. The results obtained from this study revealed that the jute plants leaf has antimicrobial properties. Therefore, jute leaf has posibility of presence of bioactive and phytochemicals compounds that might be dependent on many factors like, solvent used, method of extraction, plant material used and methodology, etc. It is proven that aqueous extract effectively inhibited the growth of the test bacteria. There are significant result but more study need to be taken to certain its antimicrobial properties with compare to over commercial antimicrobial agents. Also research should befollow up to determine the Minimum Inhibitory Concentration (MIC) and Minimum Bacteriocidal Concentration (MBC) of jute leaf of different varieties. Moreover, Activity Index (AI) and Relative Percentage Inhibition (RPI) should be estimate of effectiveness of the inhibitory action of the extract as compared to standard drug and extracts should be quantitatively analyzed for all the phytochemicals which might lead to final estimate of phytochemicals responsible for the antibacterial activity. Thus, extracts having higher AI and RPI, can be used to develop alternative drug from the plant as it is more effective on the pathogenic bacteria. Finally, the result has positive indication and has potential future which needs furthur investigation.

2.2.4.2.4.4 References

[1] Nasreen A, Ahmed Z, Ali M, Rahman T, Tahmina. Effect of processing and preservation on L-ascorbic acid content in commercially cultivated jute leaves. GSC Advan Res Rev. 2021a; 9(3): 63–69.

[2] Nasreen A, Ahmed Z, Ali M,Shimu T. Jute leaves: A potential sources of lycopene. Inter.J.Veg. Sci. 2021b; 1-9.

[3] Nasreen MA, Ahmed Z, Ali MM, Tahmina. Determination of β-carotene in jute leaves by spectrophotometry and thin layer chromatography. World J Biol Pharm Health Sci. 2022; 9(2): 11–20.

[4] Dharajiya D, Moitra N, Patel B, Patel RK. Preliminary phytochemical analysis of the Indian medicinal plants for antibacterial activity against bovine mastitis pathogens. Wayamba J Animal Sci. 2012; ISSN: 2012-578X, Number 1342590628.

[5] Iwu MW, Duncan AR, Okunji CO. New antimicrobials of plant origin, in:J. Janick (Ed.), Perspectives on new crops and new uses.ASHS Press, Alexandria. 1999; 457-462.

[6] Selvamohan T, Ramadas V, Kishore S. Antimicrobial activity of selected medicinal plants against some selected human pathogenic bacteria. Adv Appl Sci Res. 2012; 3(5): 3374-3381.

[7] Kanthimathi M, Soranam R. Antibacterial effects of Emblica officinalis and Phyllan thusniruri crude extracts against bacterial pathogens. Int J Pharm Clin Sci. 2013; 3(3): 20- 23.

[8]Dharajiya D, Patel P, Moitra N. Antibacterial activity of Emblica officinalis (Gaertn.) Fruits and Vitex negundo (L.) Leaves. Curr Tren Biotech Pharm. 2015; 9(4): 357-368.

[9] Saeed S, Tariq P. Antibacterial activities of Emblica officinalis and Coriandrum sativum against Gram negative urinary pathogens. Pak J Pharm Sci. 2007; 20(1): 32-35.

[10] Marjorie MC. Plant products as antimicrobial agents. Clin Microbiol Rev.1999; 12(4): 564-582.

[**11**] Ramkumar KM, Rajaguru P, Ananthan R. Antimicrobial properties and phytochemical constituents of an antidiabetic plant Gymnema montanum. Adv Biol Res. 2007; 1(1-2): 67-71.

[**12**] Dharajiya D, Khatrani T, Patel P, Moitra N. Evaluation of antifungal activity of Emblica officinalis, Aloe vera and Vitex negundo extracts. J Chem Bio Phy Sci. 2015; 5(4): 3990-3996.

Publisher: Eliva Press Global Ltd

Email: info@elivapress.com

Made in the USA
Monee, IL
05 November 2026

67811172R00138